LEISHMANIA

World Class Parasites

VOLUME 4

Volumes in the World Class Parasites book series are written for researchers, students and scholars who enjoy reading about excellent research on problems of global significance. Each volume focuses on a parasite, or group of parasites, that has a major impact on human health, or agricultural productivity, and against which we have no satisfactory defense. The volumes are intended to supplement more formal texts that cover taxonomy, life cycles, morphology, vector distribution, symptoms and treatment. They integrate vector, pathogen and host biology and celebrate the diversity of approach that comprises modern parasitological research.

Series Editors
Samuel J. Black, *University of Massachusetts, Amherst, MA, U.S.A.*
J. Richard Seed, *University of North Carolina, Chapel Hill, NC, U.S.A.*

LEISHMANIA

edited by

Jay P. Farrell
University of Pennsylvania.

KLUWER ACADEMIC PUBLISHERS
Boston / Dordrecht / London

Distributors for North, Central and South America:
Kluwer Academic Publishers
101 Philip Drive
Assinippi Park
Norwell, Massachusetts 02061 USA
Telephone (781) 871-6600
Fax (781) 681-9045
E-Mail <kluwer@wkap.com>

Distributors for all other countries:
Kluwer Academic Publishers Group
Post Office Box 322
3300 AH Dordrecht, THE NETHERLANDS
Telephone 31 786 576 000
Fax 31 786 576 474
E-Mail <services@wkap.nl>

 Electronic Services <http://www.wkap.nl>

Library of Congress Cataloging-in-Publication Data

Leishmania / edited by Jay P. Farrell
   p. cm. – (World class parasites ; v. 4)
   Includes bibliographical references and index.
   ISBN 1-40207-036-5
   1. Leishmaniasis. 2. Leishmania. I. Farrell, J. (Jay) II. Series.

RC153 .L45 2002
616.9'364—dc21   2002025496

TABLE OF CONTENTS

PREFACE

Parasites of the genus *Leishmania* are dimorphic protozoans that exist as intracellular amastigotes within mammalian mononucular phagocytes and as flagellated promastigotes within sand fly vectors. Although various species of *Leishmania* can infect a variety of hosts, the primary importance of these parasites is ability to produce disease in humans and dogs. Human leishmaniasis occurs in tropical, sub-tropical and temperate regions of the world with an estimated 1.5 to 2 million new cases each year. Continued research efforts have markedly increased our understanding of the ecology, epidemiology, and immunology of this disease and recent efforts have dramatically enhanced our knowledge of the cell and molecular biology of leishmanial parasites. However, we are still a long way from controlling this important disease.

This book comprises a series of chapters by authors who have devoted their research efforts to the study of *Leishmania* and leishmaniasis. Transmission and control are covered in the first four chapters with Jean-Pierre Dedet starting us off by reviewing the current worldwide epidemiology of leishmaniasis and pointing out the increasing problems associated with HIV-*Leishmania* co-infection. Next, Jeffrey Shaw discusses the incredible diversity of New World leishmanial species and the ecology of these parasites and their reservoir hosts. Robert Killick-Kendrick then takes up the biology of the sand fly and reviews methods to control transmission of infection to humans and important reservoir hosts such as the dog. Finally, Leana Campino discusses canine leishmaniasis, a topic central to understanding the transmission of human infection since dogs are important reservoir hosts for several species of *Leishmania*, especially those causing visceral disease.

The next several chapters focus on topics of a more experimental nature. The biology of promastigote development within sand flies and the biochemistry of parasite molecules that influence life in the sand fly intestine are covered by Shaden Kamhani. This has been an area of intense study since the discovery that leishmanial lipophosphoglycan mediates attachment pf promastigotes to sand fly gut epithelial cells. Another area impacted by molecular technology is reseach on the metabolic biochemistry of leishmanial parasites and Scott Landfear uses one aspect of parasite metabolism, namely the nature and regulation of membrane transporters, to discuss our understanding of the cell biology of these parasites. The interaction of *Leishmania* with host macrophages in the topic of the next two chapters. David Mosser and Andrew Brittingham focus on the interaction of parasite molecules and host cell receptors that mediate entry into macrophages and discuss how opsonization by antibody may promote the production of cytokines that promote parasite survival.

Greg Matlashewski takes a different tack and discusses how infection with different leishmanial species alters macrophage gene expression and signaling pathways. Dan Zilberstein and Moshe Ephros then cover an area critical to the control of leishmaniasis, namely chemotherapy. In addition to reviewing existing drug therapies, they discuss the current understanding of the modes of drug action and mechanisms of drug resistance.

The final three chapters are devoted to the immune response to leishmanial infections. Paul Kaye discusses the current state of knowledge of the immune response to human visceral leishmaniasis and relates it to insights gathered from studies of *L. donovani* in experimental models while I take a similar approach to cover the topic of immunity to human and murine cutaneous leishmaniasis. Finally, Antonio Campos-Neto reviews, in depth, the experimental basis of vaccine development and discusses recent efforts to move experimental vaccines from the laboratory to the field.

Leishmaniasis is a complex of infections caused by multiple leishmanial species that cause diverse forms of clinical disease. Each of these species has it own sand fly vectors and reservoir hosts and the ecological aspects that affect their transmission varies enormously. Thus, not all aspects of leishmaniasis are covered in this small volume. For example, little space is devoted to diagnosis and the clinical aspects of infection are only discussed in passing. However, these topics are detailed in multiple texts on clinical tropical medicine. Also missing is a discussion of a newly emerging area that will have enormous impact on our understanding of the biology of *Leishmania*, namely the *Leishmania* genome project. A number of laboratories are part of the *Leishmania* genome network and are actively sequencing leishmanial genes as well as pursuing the development and application of proteomics to complement these sequencing activities. As data continues to emerge from this effort, researchers will have exciting new tools to help tackle both clinical and experimental problems relevant to understanding and controlling this important infectious disease. We can only hope so since, as Professor Dedet concludes in his chapter about our current ability to control leishmaniasis, "there is much still to be done."

Jay P. Farrell, January 2002

CURRENT STATUS OF EPIDEMIOLOGY OF LEISHMANIASES

Jean-Pierre Dedet

Centre National de Référence des leishmanioses ; Laboratoire de
Parasitologie, CHU de Montpellier, 163, rue Auguste-Broussonet,
34090 Montpellier, France

INTRODUCTION

Leishmaniases are parasitic diseases caused by protozoan flagellates
of the genus *Leishmania*. Widely distributed around the world, they
threaten 350 million people in 88 countries of four continents. The annual
incidence of new cases is estimated between 1,5 and 2 million (1). In
numerous countries, increasing risk factors are making leishmaniases a
major public health problem (2).

In humans, the genus *Leishmania* is responsible for various types of
disease : visceral (VL), cutaneous (CL), of localized or diffuse type, and
mucocutaneous leishmaniases (MCL). This variability of the clinical
features results from both the diversity of the *Leishmania* species and the
immune response of its hosts.

The genus *Leishmania* includes around 30 different taxa, which
have a common epidemiology. They infect numerous mammalian species,
including humans, acting as a reservoir, and are transmitted through the
infective bite of an insect vector, the phlebotomine sandfly.

PARASITE

Leishmania are protozoa belonging to the order Kinetoplastida and
the family Trypanosomatidae. They are dimorphic parasites which present
as two principal morphological stages: the intracellular amastigote, within
the mononuclear phagocytic system of the mammalian host, and the
flagellated promastigote within the intestinal tract of the insect vector and in
culture medium.

Description

The amastigote stage is a round or oval body about 2-6 μm in
diameter, containing a nucleus, a kinetoplast and an internal flagellum seen
clearly in electron micrographs. The amastigotes multiply within the
parasitophorous vacuoles of macrophages.

The promastigote stage has a long and slender body (about 15-30 μm by 2-3 μm), with a central nucleus, a kinetoplast and a long free anterior flagellum.

Identification

Since the creation of the genus *Leishmania* by Ross (3), the number of species described has constantly increased. As the different species are indistinguishable by their morphology, other criteria have been used for their identification. Lumsden distinguished between extrinsic characters (such as clinical features, geographical distribution, behaviour in culture, laboratory animals or vectors) and intrinsic ones (such as immunological, biochemical or molecular criteria) (4). Among them, isoenzyme electrophoresis remains the current gold standard technique, while DNA-based techniques are being used increasingly.

Classification

Various types of classification have been successively applied to this genus. Those proposed between 1916 and 1987 were monothetic Linnean classifications based on few hierarchical characters. Lainson and Shaw are the authors who worked the most on these types of classification and who made them evolutive. Their last classification (5) divided the genus *Leishmania* into two sub-genera : *Leishmania* sensu stricto, present in both Old and New World, and *Viannia*, restricted to New World. Within these two sub-genera various species complexes were individualized.

Since the 1980s, Adansonian phenetic classifications have been employed. They are based on a number of similarly weighted characters (absence of hierarchy) used simultaneously (polythetic classification) without a prior hypothesis. They were at first phenetic. Isoenzymes are considered as different allelic forms of a gene, and enzymatic variation at a given locus can be interpreted as a mutation occurring during evolution. Subsequently, phylogenetic classification revealed a parental relationship between the different species of *Leishmania* (6, 7) (Table 1).

The phenetic, and particularly the cladistic classifications confirmed the majority of the taxonomic groups previously established by the Linnean classifications, and particularly that of Lainson and Shaw (5). The concordance between them mutually validated the extrinsic and intrinsic identification criteria. However, cladistic analysis allowed a more detailed study of some groups and lead to the establishment of new complexes (*L. infantum*, *L. turanica* and *L. guyanensis*), and also to the grouping in the same complex of taxa previously separated (*L. guyanensis*, *L. panamensis* and *L. shawi*).

Table 1. *Simplified classification of the genus Leishmania, derived from the phylogenetic analysis of Rioux et al. based on isoenzymes (6).*

Sub-genus <u>*Leishmania* Ross, 1903</u>

- *L. donovani* complex :   *L. donovani* (Laveran & Mesnil, 1903)

     L. archibaldi Castellani & Chalmers , 1919
- *L. infantum* complex :   *L. infantum* Nicolle, 1908

     (syn. *L. chagasi* Cunha & Chagas, 1937)
- *L. tropica* complex :   *L. tropica* (Wright, 1903)
- *L. killicki* complex :   *L. killicki* Rioux, Lanotte & Pratlong, 1986
- *L. aethiopica* complex :   *L. aethiopica* Bray, Ashford & Bray, 1973
- *L. major* complex :   *L. major* Yakimoff & Schokhor, 1914
- *L. turanica* complex :   *L. turanica* Strelkova, Peters & Evans, 1990
- *L. gerbilli* complex :   *L. gerbilli* Wang, Qu & Guan, 1964
- *L. arabica* complex :   *L. arabica* Peters, Elbihari & Evans, 1986
- *L. mexicana* complex :   *L. mexicana* Biagi, 1953

     (syn. *L. pifanoi* Medina & Romero, 1959)
- *L. amazonensis* complex :*L. amazonensis* Lainson & Shaw, 1972

     (syn. *L. garnhami*, Scorza *et al.*, 1979)

     L. aristidesi Lainson & Shaw, 1979
- *L. enriettii* complex :   *L. enriettii* Muniz & Medina, 1948
- *L. hertigi* complex :   *L. hertigi* Herrer, 1971

     L. deanei Lainson & Shaw, 1977

Sub-genus <u>*Viannia* Lainson and Shaw, 1987</u>

- *L. braziliensis* complex : *L. braziliensis* Vianna, 1911.

     L. peruviana Velez, 1913
- *L. guyanensis* complex : *L. guyanensis* Floch, 1954

     L. panamensis Lainson & Shaw, 1972

     L. shawi Lainson *et al.*, 1989
- *L. naiffi* complex :   *L. naiffi* Lainson & Shaw, 1989
- *L. lainsoni* complex :   *L. lainsoni* Silveira *et al.*, 1987

VECTOR

Sandflies are Diptera of the family Psychodidae, sub-family Phlebotominae. Their life cycle includes two different biological stages : the air flying adult and the development phase, which includes egg, four larval stages and pupa, and occurs in wet soil rich in organic material(8)

The adults are small flying insects of about 2-4 mm in length, with a yellowish hairy body. During day, they rest in dark and sheltered places (resting sites). They are active at dusk and during the night. Both sexes feed on plants, but females also need a blood meal before they are able to lay eggs. Reptiles, amphibia, birds and mammals are potential hosts. Feeding habits depend on the sandfly species, and the nature of the host on which it takes its blood-meal is a key point for *Leishmania* transmission.

About 800 species of sandflies have been described, and they are divided into five widely accepted genera : *Phlebotomus* and *Sergentomyia* in the Old World, and *Lutzomyia, Brumptomyia* and *Warileya* in the New World (9). Among these species, about 70, belonging to the genera *Phlebotomus* and *Lutzomyia,* are proven or suspected vectors of *Leishmania,* and a certain level of specificity between *Leishmania* and sandfly species exists (10)

RESERVOIR

Most leishmaniases are zoonoses and the reservoir hosts are various species of mammals which are responsible for the long term maintenance of *Leishmania* in nature (11). Depending on the focus, the reservoir can be either a wild or a domestic mammal, or even in particular cases human beings. In the case of visceral leishmaniasis, these different types of reservoir represent different evolutive steps on the path towards anthropisation of a wild zoonosis (12). Most reservoir hosts are well adapted to *Leishmania* and develop only mild infections, which may persist for many years, an important exception being the dog which commonly develops a generalized and fatal disease.

The reservoirs are included in seven different orders of mammals. Rodents, hyraxes, marsupials and edentates are reservoirs of wild zoonotic CL. Dogs are currently considered as true reservoirs of *L. infantum* and *L. peruviana*, two species which have peri-domestic or even domestic transmission. Humans are the commonly recognized reservoir host of *L. donovani* VL and *L. tropica* CL.

LIFE CYCLE

In nature, *Leishmania* are alternatively hosted by the insect (flagellated promastigote) and by mammals (intracellular amastigote stage). When a female sandfly takes a blood meal from a *Leishmania*-infected mammal, intracellular (and maybe also extracellular) amastigotes are ingested by the insect. Inside the blood meal, amastigotes transform into motile promastigotes, which escape through the peritrophic membrane enveloping the blood meal. The promastigotes multiply intensively inside the intestinal tract of the sandfly, successively as free elongated promastigotes (nectomonads) or as attached pro- and paramastigotes (haptomonads) (13). This intraluminal development occurs in the midgut (*Leishmania* sub-genus, previously section Suprapylaria according to Lainson and Shaw, 14), or in the hindgut and the midgut (*Viannia* sub-genus, previously section Peripylaria). Whatever the multiplication site, the parasites subsequently migrate to the anterior part of the sandfly midgut where they change into free-swimming metacyclic promastigotes, the stage infective for the vertebrate host.

The bite of an infected sandfly deposits infective metacyclic promastigotes into the mammalian skin, which are rapidly phagocytosed by cells of the mononuclear phagocytic system. The intracellular parasites change into amastigotes, which multiply by simple mitosis.

When the intracellular development of the amastigotes remains localized at the inoculation site, various cytokines are released and cell reactions are generated, resulting in the development of a localized lesion of CL (15). In other instances, the parasites spread to the organs of the mononuclear phagocytic system, giving rise to VL. Amastigotes may also spread to other cutaneous sites, as in diffuse cutaneous leishmaniasis (DCL), or to mucosae, in the case of MCL.

The localization of the parasite to the various organs of the patient results in the clinical expression of the disease. It is directly related to the tropism of the parasite species. In that sense, the genus *Leishmania* can be divided broadly into viscerotropic (*L. donovani, L. infantum*) and dermotropic species (roughly all other species). *L. braziliensis*, and more rarely *L. panamensis*, are known for their secondary mucosal spread. But some exceptions occur.

The clinical expression of the leishmaniases depends, not only on the genotypic potential of the different parasites, but also on the immunological status of the patient. Since the HIV infection has spread to areas where leishmaniasis is endemic, DCL has been found with species which were never previously known to cause this form of disease (e.g. *L. major* and *L. braziliensis*). In immunodeficient patients, particularly those with HIV infection, the dermotropic variants of the viscerotropic species are responsible for VL (16).

TRANSMISSION

The inoculation of metacyclic promastigotes through the sandfly bite is the usual way of leishmaniasis transmission. Other routes remain exceptional. In VL, a few cases of congenital and blood transfusion transmission have been reported. A case of direct transmission by sexual contact has been reported (17). Exchange of syringes has been incriminated to explain the high prevalence of L. infantum/HIV co-infection in intravenous drug-users populations in South Europe (18).

In CL, contact with active lesions is innocuous, infection should require inoculation of material from active sores, as was carried out in ancient times by various populations of endemic areas as a crude form of vaccination.

GEOGRAPHICAL DISTRIBUTION

Leishmaniases are widely distributed around the world. They range over the intertropical zones of America and Africa, and extend into temperate regions of South America, South Europe and Asia. Their extension limits are latitude 45° north and 32° south. Geographical distribution of the diseases is related to that of the sandfly species acting as vectors, their ecology and the conditions of internal development of the parasite.

Leishmaniases are present in 88 countries, of which 16 are industrialized and 72 developing countries, 13 of them among the poorest in the world (1).

Visceral leishmaniasis

VL is found in 47 countries and its mean annual incidence is estimated around 500 000 new cases. The main historical foci of endemic VL are located, east to west, in China, India, Central Asia, East Africa, Mediterranean basin and Brazil. The anthroponotic species L. donovani is restricted to China, India and East Africa, while the zoonotic species L. infantum extends from China to Brazil.

At the present time, 90% of the VL cases in the world are in Bangladesh, India and Nepal, Sudan and Brazil. India is certainly the biggest focus of VL in the world. Between 1875 and 1950, the disease took on an epidemic feature, with three severe outbreaks in Assam and a subsequent extension to other Indian states (19). Bihar state of India experienced a dramatic epidemic with more than 300 000 cases reported between 1977 and 1990, with a mortality rate over 2% (20). In southern Sudan, an outbreak was responsible for 100 000 deaths from 1989 to 1994, in a population of Upper Nile Province of less than one million (21). Population movements, such as rural to suburban migration in north-eastern

Brazil (22), are factors for VL extension, by exposing thousands of non-immune individuals to the risk of infection.

Cases of VL during HIV infection have regularly been recorded in various foci in the world, particularly in Southern Europe (1). For the period 1990-98, there were a total of 1,616 cases reported, 87% of which were located in the Mediterranean basin : Spain 56.7%, southern France 17.7%, Italy 14.9% and Portugal 8.1%. Cases were also reported from countries of North Africa. In the southern European countries, the prevalence of VL during HIV infection was around 2% (23, 24). The spread of AIDS to rural area where VL is endemic, and the spread of VL to suburban areas, has resulted in a progressively increasing overlap between the two diseases, not only in the Mediterranean basin, but also in the historical foci of VL, such as East Africa, India and Brazil.

Tegumentary leishmaniasis

The large majority of Old World CL are due to the two species *L. major* and *L. tropica* and proceed from countries of Near- and Middle-East : Afghanistan, Iran, Saudi Arabia and Syria. *L. major*, the species responsible for zoonotic CL, has a large distribution area, including West-, North- and East-Africa, Near- and Middle-East and Central Asia. Economic developments have been accompanied by movement of populations which caused dramatic epidemic outbreaks of this species in several countries of Middle-East, but also in Algeria and Tunisia. The anthroponotic species *L. tropica* is present in various cities of Near- and Middle-East, but extends also to Tunisia and Morocco, where an animal reservoir is suspected in some foci. Other species have restricted distribution areas : *L. aethiopica* to Ethiopia and Kenya, *L. arabica* to Saudi Arabia and *L. killicki* to Tunisia. *L. turanica* and *L. gerbilli* are Central Asian species restricted to rodents.

In the New World, *L. braziliensis* is the species which has the widest distribution area. It extends from south of Mexico to north of Argentina. *L. amazonensis* has a large distribution in South America, but human cases of this rodent enzootic species are unusual. Other species have more restricted areas : *L. guyanensis* (north of the Amazonian basin), *L. panamensis* (Colombia and Central America), *L. mexicana* (Mexico and Central America), and last *L. peruviana,* which is restricted to Andean valleys of Peru. With the exception of this later species, all other American dermotropic species are responsible for wild zoonoses within the rain forest.

PROPHYLAXIS

Intervention strategies for prevention or control are hampered by the variety of the structure of leishmaniasis foci, with many different animals able to act as reservoir hosts of zoonotic forms and a multiplicity of sandfly vectors, each with a different pattern of behaviour. In 1990, a WHO Expert Committee described no less than 11 distinct eco-epidemiological entities and defined control strategies for each of them (25).

Prevention

The aim of prevention is avoiding host infection (human or canine) and its subsequent disease. It includes means to prevent intrusion of people into natural zoonotic foci and ways to protect against infective bites of sandflies. Prevention can be at an individual level (use of repellents, pyrethroïds impregnated bed-nets, self protection insecticides), or at a collective level (indoor residual spraying, forest clearing around human settlements).

Control

Control programmes are intended to interrupt the life cycle of the parasite, to limit or, ideally, eradicate, the disease. The structure and dynamics of natural foci of leishmaniasis are so diverse that a standard control programme cannot be defined and control measures must be adapted to local situations. The strategy depends on the ecology and behaviour of the two main targets, which are not mutually exclusive : the reservoir hosts and the vectors. Control measures will be totally different depending on whether the disease is anthroponotic or zoonotic. In the New World, almost all the leishmaniases are sylvatic, and control is not usually feasible. Even removal of the forest itself may not reach the objective, as various *Leishmania* species have proved to be remarkably adaptable to environment degradation.Case detection and treatment are recommended when the reservoir is man or dog, while massive destruction may be the chosen intervention if the reservoir is a wild animal. As far as vectors are concerned, control of breeding sites is limited to the few instances where they are known (rodent burrows for *P. papatasi* and *P. duboscqi*). Antiadult measures consist of insecticide spraying.

Malaria control programmes, based on indoor residual insecticide spraying have had a side benefit for leishmaniasis incidence in several countries where a resurgence of leishmaniasis was observed in numerous areas after the ending of these campaigns : India (26), Italy, Greece and the Middle-East (27) and Peru (28).

In practice, control programmes include several integrated measures targeted not only to the reservoir and/or vector, but also associated environmental changes. Health education campaigns can considerably improve the efficiency of control programmes.

National leishmaniasis control programmes have been developed in various countries to face endemics or epidemics (India, China and Brazil for VL, Central Asia Republics of the former USSR and Tunisia for CL).

In conclusion, leishmaniases are diseases widely distributed around the world, and are an important public health problem in various countries. In spite of the progress in understanding most facets of their epidemiology, control of leishmaniases remains unsatisfactory. Control is hampered by the wide variety of structures of foci, with many different animals able to act as reservoirs hosts of zoonotic forms and a multiplicity of sandfly vectors, each with a different pattern of behaviour. There is much still to be done.

REFERENCES

1.Desjeux, P. 1999. Global control and *Leishmania*/HIV co-infection. Clin Dermatol 17:317-325.

2.Desjeux, P. 2001. The increase in risk factors for leishmaniosis worldwide. Trans R Soc Trop Med Hyg 95:239-243.

3.Ross, R. 1903. Note on the bodies recently described by Leishman and Donovan. Brit Med J 2:1261-1262.

4.Lumsden, WHR. 1974. Biochemical taxonomy of *Leishmania*. Trans R Soc Trop Med Hyg 68:74-75.

5.Lainson, R and Shaw, JJ. Evolution, classification and geographical distribution. In: Peters W, Killick-Kendrick R, eds. 1987. The leishmaniases in Biology and Medicine. London. Academic Press 1:1-120.

6.Rioux, JA, Lanott,e G, Serres, E, et al. 1990.Taxonomy of *Leishmania*. Use of isoenzymes. Suggestions for a new classification. Ann Parasitol hum comp. 65:111-125.

7.Rioux, JA and Lanotte G. 1993. Apport de la cladistique à l'analyse du genre *Leishmania* Ross, 1903 (Kinetoplastida, Trypanosomatidae). Corollaires épidémiologiques. Biosystema. 8:79 90.

8.Ward, RD. 1985. Vector biology and control. In : Chang KP, Bray RS eds. Leishmaniasis.Amsterdam: Elsevier199-212.

9.Lewis, DJ. 1974. The biology of Phlebotomidae in relation to leishmaniasis. Ann Rev Entomol 19:363-384.

10.Killick-Kendrick, R. 1990. Phlebotomine vectors of the leishmaniases: a review. Med Vet Entomol 4: 1-24.

11.Ashford, RW. 1996. Leishmaniasis reservoirs and their significance in control. Clin Dermatol 14: 523-532.

12.Garnham, PCC. 1965. The Leishmanias, with special reference of the role of animal reservoirs. Am Zool 5:141-151.

13.Walters, LL. 1993. *Leishmania* differentiation in natural and unnatural sandfly hosts. J Euk Microbiol 40:196-206.

14.Lainson, R and Shaw, JJ. 1979. The role of animals in the epidemiology of south American leishmaniasis. In : Lumsden WHR, Evans DA, eds. Biology of the Kinetoplastida London: Academic Press 2:1-116.

15.Barral-Netto, M, Machado, P and Barral, A. 1995. Human cutaneous leishmaniasis : recent advances in physiopathology and treatment. Europ J Dermatol 5:104-113.

16.Pratlong, F, Dedet, JP, Marty, P, et al. 1995. *Leishmania*-human immunodeficiency virus coinfection in the Mediterranean Basin : isoenzymatic characterization of 100 isolates of the *Leishmania infantum* complex. J Infec Dis 172:323-326.

17.Symmers, WSC. 1960. Leishmaniasis acquired by contagion. A case of marital infection in Britain. Lancet 1:127-132.

18.Alvar, J and Jimenez, M. 1994. Could infected drug-users be potential *Leishmania infantum* reservoirs ? AIDS 8:854.

19.Dye, C and Wolpert, DM. 1988. Earthquakes, influenza and cycles of Indian kala-azar. Trans R Soc Trop Med Hyg 82:843-850.

20.Thakur, CP and Kumar, K. 1992. Post kala-azar dermal leishmaniasis : a neglected aspect of kala-azar control programmes. Ann Trop Med Parasitol 86:355-359.

21.Seaman, J, Ashford, RW, Schorscher, J, et al.1992. Visceral leishmaniasis in southern Sudan: status of healthy villagers in epidemic conditions. Ann Trop Med Parasitol 86:481-486.

22.Badaro, R, Jones, TC, Carvalho, EM, et al. 1986. New perspectives on a subclinical form of visceral leishmaniasis. J infect Dis 154:1003-1011.

23.Marty, P, Le Fichoux, Y, Pratlong, F, et al. 1994. Human visceral leishmaniasis in Alpes Maritimes, France: epidemiological characteristics for the period 1985-1992. Trans R Soc Trop Med Hyg 88:33-34.

24.Alvar, J. 1994. Leishmaniasis and AIDS co-infection: the Spanish example. Parasitol Today 10:160-163.

25.World Health Organization. Control of the leishmaniases. Report of a WHO Expert Committee. 1990. Technical Report Series 793:173.

26.Sanyal, RK, Banjerjeeb, DP, Ghosh, TK, et al. 1979. A longitudinal review of kala-azar in Bihar. J Commun Dis 11:149-169.

27.Saf'janova, VM. 1971. Leishmaniasis control. Bull Wld Hlth Organ 44:561-566.

28.Davie,s CR, Llanos-Cuentas, A, Canale,s J, et al. 1994. The fall and rise of Andean cutaneous leishmaniasis : transient impact of the DDT campaign in Peru. Trans R Soc Trop Med Hyg 88:389-393.

NEW WORLD LEISHMANIASIS: THE ECOLOGY OF LEISHMANIASIS AND THE DIVERSITY OF LEISHMANIAL SPECIES IN CENTRAL AND SOUTH AMERICA.

Jeffrey J. Shaw

Departamento de Parasitologia, Instituto de Ciências Biomédicas, Universidade de São Paulo, São Paulo, Brazil.

INTRODUCTION

The vertebrate hosts of neotropical *Leishmania* range from rodents, edentates, carnivores to non-human primates (1-4). Their vectors, phlebotomine sand flies (Diptera: Psychodidae), have managed to occupy practically every ecological niche that occurs in Central and South America, ranging from the semi-arid mountainous regions of the Andes to the luxuriant forest of Amazonia. Because of this there are very few habitats in which *Leishmania* species are not being transmitted. There are some 400 sand fly species (5). Of these, approximately 30 (6) are either suspected or proven vectors. The number of species involved as reservoirs and vectors is, in the author's opinion, a gross underestimation. It is against this backcloth of enormous biological diversity that the neotropical *Leishmania* are speciating. Ecological changes that effected the distribution of vectors and reservoirs were natural phenomena that have occurred during the past 80 million years - the estimated time span of evolution (7, 8) from a single ancestor (9) that begun with a bifurcation into two major groups (10) known today as the subgenera *L.(Leishmania)* and *L.(Viannia)* (2). Those of the former subgenus appear to have the Old World as their ancestral home while the latter subgenus evolved in the Americas and never managed to get to the Old World. The presence of the subgenus *L.(Leishmania)* in the Americas is thought to represent a second influx from the Old World after the division of the continents. Another possibility (11) is that the whole genus evolved in the New World and then spread to the Old World.

The New World species evolved when their hosts were exposed to periods of global warming and cooling that resulted in the expansion and contraction of the forests, semi-arid regions and deserts (PROJECT–PALEOVEGETATION DATASET http://www.soton.ac.uk/~tjms/nerc.html). Minor climatic changes, such as today's "El Niño", were most probably superimposed upon these major climatic periods, again creating ecological changes which effected both vectors and hosts. To help us understand today's taxonomic diversity of the *Leishmania* we need to try to envisage the kind of population structure of the ancestral *Leishmania*. Clearly we can only guess at what it was but the realization that today's populations have a clonal structure (12) gives us a good working hypothesis.

Past and present environmental changes affect both vector and host and their relationship to each other. These ever changing ecological pressures are with no doubt the driving force that resulted in the mosaic of leishmanial genetic variation that is presently being detected with molecular tools.

In this chapter the named species will be listed with brief details of their life cycles in an ecological context. Without a thorough knowledge of the ecology of each *Leishmania* species it is impossible to say whether or not it is feasible or even desirable to attempt to control them or evaluate the efficiency of control programs.

ECOLOGICAL CHANGES ASSOCIATED WITH MAN'S ACTIONS.

One of the principal phenomena that changed the ecology of leishmaniasis in recent times is man himself undoubtedly. Deforestation has been extensive throughout Latin America and has changed the composition of the mammalian and sand fly fauna. This determines in turn what *Leishmania* species survives. There must be a minimal vector/reservoir ratio and habitat size below which transmission is unviable but we have no idea what this might be. Some years ago it was predicted that leishmaniasis would disappear from the São Paulo State since the forest had been almost completely destroyed (13) and there were no reports of disease in man. However, human cases reappeared indicating that both vectors and reservoir hosts had adapted to a modified environment that today consists principally of non-climax forest.

We can see at first hand how deforestation completely changes an area. However, other environmental changes are being discovered such as global warming and the reduction of the ozone layer. It is questionable whether these mimic natural phenomena of the past but the speed at which they occur is unprecedented. In Amazonia forests fires are occurring naturally (14), a thing that was unheard of in the living memory of its inhabitants.

Man's dwellings have been evolving in various forms in Latin America ever since he first colonized Central and South America, some 12,000 years ago. One might hypothesize that one reason the Inca civilization built its cities so high was to escape sand flies and other haematophagous insects. Both man and domestic animals are excellent blood sources for them and over the years some species became adapted to this relatively new habitat, which for many years was primarily set in rural areas. *Lu .longipalpis* **sensu stricto** and *Lu. intermedia* are excellent examples of sand flies that adapted to this habitat. However, sand flies have more recently invaded the high rises of suburbia. In Belo Horizonte, the capital of Minas Gerais State, Brazil, *Lu. longipalpis* and dogs infected with visceral leishmaniasis have been found in the centre of the city (15), thus divorcing themselves from their rural origins. However, we must be careful when interpreting the use of the word urban. Some places that are described as being urban in publications are peripheral regions of small towns or small villages in which there are continuous bands of vegetation. Others, however, such as the more central regions of Belo Horizonte, are fully fledged urban areas that have no rural flavor with only small patches of non-continuous vegetation.

There are, however, some surprising facts that show that man's actions are not always destructive. Palaeoecological studies (16) have shown that human disturbances changed the ecology of the Amazonia some 6500 years ago. The forest coverage remained but there were changes in its composition. There was an intensification of agriculture some 3500 years ago which recently terminated in a period of regeneration, probably linked to the arrival of the first Europeans in Amazonia.

THE *LEISHMANIA* SPECIES AND THEIR ECOLOGY.

The distribution of the *Leishmania* species in Central and South America is given in Table I in which man serves as an excellent sentinel animal for mapping the distribution of the different *Leishmania*. In many countries the presence of a *Leishmania* species is only known from man and there are no records from sand flies or animal reservoirs The country code of the International Code Number (17), used to identify *Leishmania* strains (www.bdt.org.br/leishnet/), is placed in brackets after the vectors and reservoirs to indicate which country the record refers to. Brief ecological profiles accompany each species. Host and vector records are taken principally from reviews (1-4). In the space available it is not possible to give a detailed country by country ecology of each *Leishmania* species. Such details can be found in the original papers and review articles.

Seasonal variations of many vector species are not well studied. There are large population fluctuations between the wet and dry seasons that effect transmission. The seasons do not always coincide with each other throughout Latin America and within a region different soils and drainage cause variations in micro habitats. Thus within very small areas such factors may result in uneven population densities of the same sand fly species.

Similarly the period of reproduction and distribution of reservoir hosts are linked to climatic conditions which also effect their availability as food sources for the vector(s).

Table I. *The neotropical Leishmania species found [1] in Central and South America and their pathologies in man.*

Country (Code)	*Leishmania* Species	Pathology[2]
Argentina (AR)	*L.(L.) infantum chagasi*	VL
	L.(V.) braziliensis	CL
Belize (BZ)	*L.(L.) mexicana*	CL
	L.(V.) braziliensis	CL
Bolivia (BO)	*L.(L.) amazonensis*	CL, ADCL
	L.(L.) infantum chagasi	VL
	L.(L.) lainsoni [a]	CL
	L.(V.) braziliensis	CL, MCL
Brazil(BR) ADCL, MCL[3], VL[4]	*L.(L.) amazonensis*	CL,
	L.(L.) forattinii	No Records[8]
	L.(L.) infantum chagasi	VL, (CL[5])
	L.(L.) major-like	CL[6]
	L.(V.) braziliensis	CL,MCL
	L.(V.) guyanensis	CL,MCL
	L.(V.) lainsoni	CL
	L.(V.) naiffi	CL
	L.(V.) shawi	CL
	L. deanei	No Records
	L .enriettii	No Records
Colombia (CO)	*L.(L.) amazonensis*	CL, ADCL
	L.(L.) infantum chagasi	VL
	L.(L.) mexicana	CL, ADCL
	L.(V.) braziliensis	CL, MCL
	L.(V.) guyanensis	CL
	L.(V.) panamensis	CL, MCL
	L. colombiensis [b]	CL
Costa Rica (CR)	*L.(L.) mexicana*	CL
	L.(V.) braziliensis	CL, MCL
	L.(V.) panamensis	CL

Dominican Rep. (DO)	*L.(L.) mexicana*-like	CL, ADCL
Ecuador (EC)	*L.(L.) major*-like [c]	CL
	L.(L.) mexicana [c]	CL
	L.(V.) braziliensis	CL, MCL
	L . equatorensis	No Records
El Salvador (SV)	*L.(L.) infantum chagasi*	VL
	L.(L.) mexicana	CL
French Guyana (GY)	*L.(L.) amazonensis*	CL, ADCL
	L.(V.) braziliensis [d]	CL, MCL
	L.(V.) guyanensis	CL
	L.(V.) naiffi [e]	CL
Guadeloupe (GP)	*L.(L.) infantum chagasi*	VL
Guatemala (GT)	*L.(L.) infantum chagasi*	VL
	L.(L.) mexicana	CL
	L.(V.) braziliensis	CL
Guyana (GY)	*L.(V.) guyanensis*	CL
Honduras (HN)	*L.(L.) infantum chagasi*	VL, CL [7]
	L.(L.) mexicana	CL, ADCL
	L.(V.) braziliensis	CL, MCL
	L.(V.) panamensis	CL, MCL
Martinique (ML)	*L.(L.)* sp.	CL
Mexico (MX)	*L.(L.) infantum chagasi*	VL
	L.(L.) mexicana	CL, ADCL, VL [m]
Nicaragua (NI)	*L.(L.) infantum chagasi*	CL, VL
	L.(V.) braziliensis	CL, MCL
	L.(V.) panamensis	CL, MCL
	L.(V.) braziliensis/panamensis [f]	CL
Panama (PA)	*L.(L.) aristedesi*	CL [n]
	L.(L.) mexicana	CL [n]
	L.(V.) braziliensis	CL
	L.(V.) panamensis	CL
	L. hertigi	No Records
	L. colombiensis [b]	No Records
Paraguay (PY)	*L.(L.) amazonensis*	CL, ADCL
	L.(L.) infantum chagasi	VL
	L.(L.) major-like [g]	CL
Peru (PE)	*L.(L.) amazonensis* [h]	ADCL

	L.(V.) braziliensis	CL, MCL
	L.(V.) guyanensis [i]	CL
	L.(V.) lainsoni [i]	CL
	L.(V.) peruviana	CL
	L.(V.) braziliensis/L.(V.) peruviana [j]	CL, MCL
	L. colombiensis [k]	CL
Surinam (SR)	*Leishmania* sp.	CL
Trinidad (TR)	*Leishmania* sp.	CL
Venezuela (VE)	*L.(L.) infantum chagasi*	VL
	L.(L.) garnhami	CL
	L.(L.) pifanoi	CL, ADCL
	L.(L.) venezuelensis	CL
	L.(V.) braziliensis	CL, MCL
	L.(V.) braziliensis/L.(V.) guyanensis [l]	CL
	L colombiensis	VL [9]

[1] The data in this table is taken from the review articles (1-4). except [a] Martinez et al. (18), [b] Kreutzer et al. (19), [c] Katakura *et al.*(20), [d] Raccurt *et al.* (21), [e] Darie *et al.*(22), [f] Darce *et al.*(23), [g] Yamasaki et al.(24), [h] Franke et al.(25), [i] Lucas et al.(26), [j] Dujadin *et al.*(27), [k] Franke Shaw, & Ishikawa (unpublished observations), [l] Bonfante *et al.*(28), [m] Monroy-Ostria et al (29), [n] Christensen et al (30)

[2] CL = Cutaneous Leishmaniasis, ADCL = Anergid Diffuse Cutaneous Leishmaniasis, MCL= Mucocutaneous leishmaniasis, VL= Visceral Leishmaniasis; [3] In cases of ADCL;

[4] Only recorded in a small region of Bahia State, BR; [5] Only in Rio de Janeiro, BR;

[6] Localized in Minas Gerais, BR; [7] Ponce *et al.*(31); [8] Human infections recorded as *L.(L.) amazonensis* from southern Brazil may be this parasite, [9] bone marrow only.

Species Of The Subgenus *Leishmania* (*Leishmania*) Ross 1903

Members of this subgenus are clearly defined by many characters as a major monophylectic evolutionary line that includes both Old and New World species. They have typically hypopylarian development in the sand fly. The parasites that have as their principal hosts rodents can be divided into the *mexicana* complex and the *amazonensis* complex which are defined by their rDNA (32). The *mexicana* complex is distributed throughout Central America, the northern coastal regions of Venezuela and some Caribbean Islands. Caribbean and Venezuelan species are in general difficult to culture and produce small lesions in hamsters. The *amazonensis* complex is found throughout South America. There appears to be some overlap between the two complexes in Panama and some Andean regions.

Leishmania of the *donovani* complex
L. (L.) infantum chagasi Cunha & Chagas 1937

Host(s): Primary peridomestic reservoirs - dogs *Canis familiaris* (BR, CO, BO, VE): wild reservoirs - crab eating foxes *Cerdocyon thous* (BR, VE), opossums *Didelphis marsupialis* (CO-(33), BR, VE(34)), spiny rats *Proechimys canicollis*(CO), domestic rats *Rattus rattus*(VE (34)). *Dusicyon vetulus* was recorded as a reservoir of *L.(L.)i.chagasi* but it is now considered that the specimens were in fact *C.thous* (35).

Vector(s): *Lutzomyia (Lutzomyia) longipalpis*(BO,BR,CO,VE). .Present in all countries except Chile where it is considered as the vector without identification of infections. There is no doubt that this is a complex composed of species and subspecies (36, 37). *Lu. evansi* (CO (38),VE (34) and *Lu. cruzi* (BR (39)) are also considered to be vectors.

Ecology: Transmission takes place in many habitats ranging from forested regions, where it occurs amongst foxes, to the peridomestic habitat of both rural and urban areas, where dogs are the principal reservoirs. *Lu. longipalpis* is found in small numbers in climax rain-forest that is distant from human dwellings and in more open areas associated with outcrops of rock. Settlers generally construct animal shelters, especially for their chickens, which are ideal resting sites for the sand fly. It only needs an infected fox or dog to enter this environment to establish a new *L. (L.)i.chagasi* focus. In Colombia *Lu. evansi* is a vector in the dry tropical forest where 5/22 opossums have been found infected. In Corumba, Mato Grosso de Sul State Brazil, *Lu.cruzi*, a species closely related to *Lu. longipalpis,* has been found infected with *L. (L.)i.chagasi* and is thus a potential vector. The present concern today is with the urbanization of the transmission cycle which is linked to the adaptation of *Lu. longipalpis* to new environments.

Transmission level: terrestrial enzootic and zoonotic cycles.

Leishmania of the *mexicana* complex.
L. (L.) mexicana Biagi, 1953 *emend* Garnham, 1962

Hosts: Forest rodents: *Ototylomys phyllotis*, (BZ, MX) *Nyctomys sumichrasti*(BZ), *Heteromys desmarestianus*(BZ), *Sigmodon hispidus* (BZ, MX(40)). *Peromyscus yucatanicus* (MX(40)), *Oryzomys melanotis*(MX(40)). Accidental host - dog (EC (41))

Vector(s): Principal vector *Lu. olmeca olmeca* (BZ, MX); secondary vectors *Lu. cruciata* BZ, MX), *Lu. ovallesi*(MX), *Lu. ylephiletor* (GT). Outside enzootic region of Central America - *Lu. ayacuchensis* (EC (41))

Ecology: This was the first neotropical *Leishmania* causing dermal leishmaniasis in man to be found in wild mammals. However, there are differences of opinion as to which rodent species are its principal hosts. In Belize *Ototylomys phyllotis* is considered as the primary wild host while in Mexican State of Campeche it is considered to be *Peromyscus yucatanicus*. It is likely that any rodent could become the principal reservoir depending on the ecological conditions that favor one or another species. However, the higher rates of infection and the longevity of the infection recorded in both *Ototylomys phyllotis* and *Peromyscus yucatanicus* (42) suggest that these rodents are the principal hosts of *L. (L.) mexicana*. There is no doubt though that the highly rodentophilic sand fly, *Lu. o. olmeca*, is the principal vector in the enzootic cycle. There are records of *Lu. o. olmeca* in Honduras, Nicaragua and Costa Rica (5) where it is almost certainly the vector. In the highlands of Ecuador *Lu. ayacuchensis* (41) has been incriminated as a vector of a parasite identified as *L. (L.) mexicana*. This is well beyond the geographical range of the *mexicana* complex(32).

Transmission level: zoonotic and enzootic cycles - ground level

L. (L.) pifanoi Medina & Romero, 1959 *emend* Medina & Romero, 1962

Host(s): There are no definite isolations of this parasite from wild animals. Infections recorded in *Heteromys anomalus* (VE), *Zygodontomys microtinus* (VE) from the Venezuela States of Carabobo and Barinas may have been this parasite.

Vector(s): a record of this parasite in *Lu.flaviscutellata* from Venezuelan Amazonia was most probably not *L. (L.) pifanoi* but *L. (L.) amazonensis*. *Lu. townsendi* (VE) has been experimentally infected with this parasite but there are no grounds for considering it to be a vector.

Ecology: This species belongs to a group of parasites of the *mexicana* complex that do not grow well in culture or hamsters. These biological characters have probably hindered advances on the ecology of this parasite.

Transmission level: unknown but probably ground level

L. (L.) venezuelensis Bonfante-Garrido, 1980

Host(s): So far only found in horses (VE) and cats (VE) which are secondary domestic hosts.

Vector(s): Isolations have been made from *Lu. rangeliana*(VE) and 3 other species (*Lu. lichyi, Lu. olmeca bicolor, Psychodopygus panamensis*) are incriminated in Venezuela as vectors on circumstantial evidence.

Ecology: This parasite belongs to the *mexicana* complex and is the dominant *Leishmania* of the subgenus *L.(Leishmania)* causing cutaneous leishmaniasis in Lara State, Venezuela. Transmission to man occurs in many different habitats suggesting that the wild reservoirs are common small animals, probably rodents and that in fact more than one vector could be involved.

Transmission level: probably ground level

L. (L.) sp Dominican Republic

Host(s): Unknown. This is an island species with unique characters that is associated with a diffuse form of cutaneous leishmaniasis

Vector(s): There are only two sand flies on the island, *Lu.christophei & Lu.cayennensis.* The former is consider as the most likely vector since it is attracted to mammals.

Ecology: Again there are only a few native mammal species and it is assumed that one or more of them must be the reservoir although no isolations have been made. It is presumed that this is a simple cycle involving one sand fly species and local mammals.

Transmission level: probably ground level

Leishmania of the *amazonensis* complex.
L. (L.) amazonensis Lainson & Shaw, 1972

Host(s): Rodents: *Proechimys*(BR,GY,TR), *Oryzomys*(BR,TR) *Neacomys*(BR), *Nectomys*(BR), *Dasyprocta* (BR) *Heteromys anomolus*(TR); marsupials: *Caluromys philander*(TR), *Didelphis*(BR,GY), *Marmosa*(BR,TR), *Metachirus*(BR), *Philander* (BR); canids *Cerdocyon* (BR). In the Pacific coastal region of Ecuador single isolates from *Sciurus vulgaris*(EC), *Potos flavus*(EC), and *Tamandua tetradactyla*(EC) were identified (43) as *L. (L.) amazonensis*.

Vector(s): *Lu. flaviscutellata* (BR,GY,TR) and probably in Colombia and Ecuador. *Lu. olmeca nociva*(BR), *Lu.reducta* (BR), *Lu. nuneztovari anglesi* (BO) (44).

Ecology: Both the principal vector and the known reservoirs occur in both climax and non-climax forests. The enzootic cycle thus occurs in many different woodland environments including plantations established in areas that were cut and burnt. However, there is no evidence of any peridomestic transmission. The biology of *Lu. flaviscutellata* is complex and population density depends on the levels of soil water. The highest numbers are associated with very humid soils, especially near small streams, the bases of earth damns and ponds or lakes whose levels are relative stable. This sand fly does not seem to be able to tolerate continual flooding and it is absent in the varzea forests. *Lu. flaviscutellata* is a catholic feeder as indicated by the variety of different ground loving mammals that have been found infected. Females feed preferentially on rodents, especially those belonging to the genus *Proechimys*.

Transmission level: ground level. *Lu. flaviscutellata* is a seldom taken above 1.5 m from the forest floor which favours intense lower story transmission. It is probably because of this behaviour that *L. (L.) amazonensis* is unknown in arboreal animals.

L. (L.) aristidesi Lainson & Shaw, 1979

Host(s): rodents - *Oryzomys capito* (PA), *Proechimys semispinosus* (PA), *Dasyprocta punctata*(PA); marsupials - *Marmosa robinsoni* (PA).

Vector(s): suspected *Lu.olmeca bicolor* (PA).

Ecology: Linked to forests where the suspected vector is common and is the dominant sand fly attracted to both rodents and marsupials. There is evidence that the rodent population is higher in interface growth between forest and farmland. The suspected vector is commoner in mature hilltop forest than in mature or secondary lowland forest. It is likely that other small mammals are involved in the enzootic cycle but that one will be dominant.

Transmission level: ground level

L. (L.) garnhami Scorza, Valera, de Scorza, Carnevali, Moreno &Lugo-Hernandez, 1979

Host(s): the opossum *Didelphis marsupials* (VE).

Vector(s): *Lu. youngi* (VE). An isolate was made from a fly identified as *Lu.townsendi* (VE) (= *Lu.youngi*) but it was not critically identified.

Ecology: So far records of this species are limited to rural and suburban habitats of the Venezuelan Andean highlands of Merida State between 800 and 1800m. The suspected vector, *Lu. youngi,* is both exo- and endophilic suggesting that a mammal with similar habits may be involved in the enzootic and zoonotic cycle. Since all members of the *amazonensis* complex are closely associated with rodents it seems likely that they may also be reservoirs of this parasite.

Transmission level: zoonotic, enzootic terrestrial.

L. (L.) forattinii Yoshida, Pacheco, Cupolillo, Tavares, Machado, Momen, Grimaldi Jr, 1993

Host(s): *Didelphis marsupialis auratus*(BR), *Proechimys iheringi denigratus* (BR)

Vector(s): *Ps. ayrozai* and *Lu.yuilli* suspected on experimental grounds

Ecology: This is a species limited to eastern Brazil including the Atlantic rainforest. Published records from man of *L.(L.)amazonensis* in these areas are probably *L. (L.) forattinii*. In the general absence of the *olmeca/flaviscutella* complex in this region it is likely that other species that are catholic feeders of small mammals are vectors but they remain to be identified.

Transmission level: ground level

Leishmania major -like parasites

Host(s): unknown

Vector(s): unknown

Ecology: *L.major*-like parasites have been recorded from man in Brazil, Ecuador and Paraguay. In some cases they were confirmed as being imported from the middle east but in others they were not. The origin of the autochthonous cases is unknown. It is possible that they represent a neotropical parasite or an imported one that has somehow established an indigenous transmission cycle.

Transmission level: zoonotic ground level

Species Of The Subgenus *Leishmania* (*Viannia*) Lainson & Shaw, 1987.

All members of the subgenus are limited to the neotropics and are clearly defined by many characters as a major monophyletic evolutionary line. They have a peripylarian development in the sand fly. Their principal hosts are Xenathra and rodents. They are important parasites of man in which zoonotic transmission is at ground level. The enzootic cycles are either arboreal or terrestrial. In the case of the former the major reservoirs are Xenathra, especially two-toed sloths. Domestic animals are infected but their role as sources of infection to sand flies in the peridomesitc environment needs to be verified.

Leishmania (*Viannia*) *braziliensis* Viannia, 1911 *emend* Matta, 1916. Subgenus Type Species.

Host(s): isolated from dogs (BR,CO,VE) donkeys, *Equus asinus* (CO, VE), horses, *Equus caballus*(BR, VE) and the rodents *Akodon arviculoides* (BR), *Bolomys lasiurus* (BR), *R.rattus* (BR)(45). By molecular methods(45) *Akodon arviculoides*(BR), *Holochilus scieurus* (BR) *Nectomys squamipes* (BR), *Didelphis albiventris*(BR), *Marmosa* sp.(BR), *R. rattus* (BR) and possibly(46) the opossums *Micoureus demerarae*(CO), *D.marsupialis*(CO) the rodents *Melanomys caliginosus*(CO), *Microryzomys minutus*(CO), *R. rattus*(CO), and the rabbit *Sylvilagus brasiliensis*(CO).

Vector(s): Parasites isolated and identified from: *Ps.ayrozai* (BO), *Ps.carrerai* (BR,BO), *Ps.complexus* (BR), *Lu. gomezi* (VE(47)), *Lu.migonei* (BR), *Ps. nuneztovari* (BO), *Lu. ovallesi* (GT(48),VE,(47)), *Lu. spinicrassa* (CO,VE), *Ps.panamensis* (GT,VE), *Lu.ylephiletor*(GT), *Ps.yucamensis* (BO) *Ps.wellcomei*(BR), *Lu,.whitmani* (BR). Suspected on epidemiological grounds *Ps.amazonensis* (BR), *Lu. intermedia* (BR), *Lu. lonigfusca* (CO), *Ps.paraensis* (BR), *Lu.pessoai* (BR), *Lu.torvida* (CO). Suspected by molecular methods(49) *Lu.pessoai* (BR), *Lu.misionensis*(BR).

Ecology: This species has been recorded from Mexico to Argentina (see Table 1). Besides domestic animals (dogs and equines) **bone fide** isolates have only been made from 3 wild mammal genera (*Akodon, Bolomys, Rattus*) and 12 vector species. Using a combination of molecular and parasitological methods different rodents (*Bolomys lasiurus, R..rattus, Nectomys)* and marsupials (*Didelphis* sp., *Marmosa* sp.) have recently been indicated as the principal reservoirs in Pernambuco State, Brazil (45). It remains to be seen if this type of reservoir mosaic is true for the rest of Latin America. The importance of small mammals as reservoirs, the extensive genetic polymorphism of this species (50) and clear geographical clustering (51) indicate a multeity of epidemiologies. The general picture is that silvatic transmission in Amazonia is limited to climax forest where *Psychodopygyus* sand flies are the vector. Beyond this region in climax, non-climax forest and even urban localities species of *Nyssomyia* are the principal vectors. There is the possibility that small mammals serve as links between the silvatic and peridomestic environments. In the latter it has been suggested that domestic animals and even man could be sources of infection to sand flies but further data is required to support this hypothesis.

Transmission level: zoonotic & enzootic terrestrial

L. (V.) peruviana Velez, 1913

Host(s): isolated from dogs(PE), opossums *Didelphis albiventris* (PE) (1/20), leaf eared mice, *Phyllotis andinum* (PE) (2/154). Positive PCRs for *L. (Viannia)* species from the tissues of *D. albiventris* (PE) and two rodents(52) *Akodon* sp., (PE), *P.andinum* (PE).

Vector(s): *Lu. peruensis* (PE), *Lu. verrucarum*(PE), potentially *Lu.noguchi* (PE).

Ecology: This species is found in dry, mountainous, rocky terrain of the Pacific watershed of the Peruvian Andes above 1200 m and may occur in the same environment in Argentina and Ecuador. Transmission appears to be greater during the summer and autumn. Although dogs and wild animals are implicated as reservoirs it is not clear which, if any, serve as the source of infection for the sand flies that transmit to man. There was a statistical connection between *Lu.noguchi* abundance and infections in man but none between human cases and dogs. The association of this sand fly with the leaf eared mouse, from which *L.(V.) peruviana* has been isolated, supports the idea of a zoonotic cycle involving these two animals. PCR results have indicated that the prevalence in dogs and wild animals is similar suggesting that there could be an enzootic cycle amongst marsupials and rodents that spans silvatic and peridomestic habitats.

Transmission level: terrestrial.

L. (V.) guyanensis Floch 1954

Host(s): Principal reservoir two-toed sloth, *Choloepus* didactylus (BR, GY); secondary reservoirs lesser anteaters, *Tamandua tetradactyla*(BR), opossums *D.marsupialis* (BR, GY)spiny rats, *Proechimys* sp. (BR,GY).

Vector(s):_Lu.umbratilis_(BR,CO,GY) and secondarily *Lu. anduzei* and *Lu. whitmani*.

Ecology: This *Leishmania* is distributed to the north of the Amazon river from Amapa and French Guiana in the east to Colombia and Ecuador in the west. The enzootic cycle in climax forest occurs at night time in the canopy when the humidity is high and temperatures are lower. During the daytime humidity falls, the temperature rises and the vector, *Lu.umbratilis*, moves down the trunks of the larger trees. *Lu.umbratilis* is very common on tree trunks during the day and this could represent a lekking behavior. The risk of transmission to man during daylight hours at the bases of these trees is very high. The infection rate of *L. (V.) guyanensis* in *D.marsupialis* is high in the Manaus area, Amazonas State, Brazil. It is postulated that the clearance of primary forest and its remnants in the valleys near to dwellings favours a cycle which excludes sloths.

Transmission level: enzootic principally arboreal; zoonotic terrestrial.

L. (V.) panamensis Lainson & Shaw, 1972

Host(s): principal reservoir two-toed sloths - *Choloepus hoffmannii*(CR,CO,PA); secondary reservoirs three-toed sloths *Bradypus infuscatus* (CR,EC,PA), monkeys - *Aotus trivigatus* (PA), *Sanguinus geoffroyi* (PA); procyonids - *Bassaricyon gabbi* (PA) *Nasua nasua* (PA), *Potus flavus* (EC,PA); rodents - *Akodon*(CO), *Coendou sp*(CO), *Heteromys demarestianus* (CR), *R..rattus*(CO); marsupials - *Metachirus nudicaudatus*(CO), *Didelphis marsupialis*(CO); dogs - *Canis familiaris*(CO).

Vector(s): *Lu.trapidoi*(CO,EC,PA), *Lu.gomezi*(PA), *Lu.panamensis* (PA), *Lu.ylephiletor* (CR,PA) and possibly *Ps.carrerai thula*= *Ps.pessoana*(PA) and *Lu.hartmanni*(EC).

Ecology: Although many arboreal mammal species are infected with this *Leishmania* the high infection rates in two-toed sloths indicate them as the primary reservoirs. It is, however, unknown if this parasite exists in regions were two-toed sloths do not occur. The primary enzootic cycle involves the canopy-loving sand fly *Lu.trapidoi* and the two-toed sloth. Sloths do come to the ground and it is conceivable that they infect species found in the lower stories such as *Lu.panamensis* and *Lu.gomezi*. Similarly infected *Lu.trapidoi* that descend to low levels in the forest could infect mammals that are less arboreal, including of course man. This fits with the overall picture of infection being commoner in arboreal animals but present in some terrestrial ones. A secondary cycle involving both mammals and sand flies that are found closer to the ground cannot, however, be ruled out. In areas where there is a *Lu.gomezi/Lu.panamensis* association human infections did not occur suggesting that alone they cannot support the enzootic cycle. Infections in dogs like man appear to be accidental.

Transmission level: enzootic principally arboreal; zoonotic terrestrial.

L. (V.) lainsoni Silveira, Shaw, Braga & Ishikawa, 1987

Host(s): *Agouti paca*(BR)

Vector(s): *Lu.ubiquitalis* (BR), possibly *Lu.velascoi* (PE)

Ecology: This is one of the few *Leishmania* that has a one vector/one reservoir cycle. Human infections now increase the geographical range of this parasite to Peru and Bolivia. It probably occurs in all the other countries wHere the vector and reservoir host occur. In the Andean region of Peru, where one human case was recorded, flagellates were seen but not identified in *Lu.velascoi* which belongs to the same subgenus as *Lu.ubiquitalis*. This raises the possibility that in some regions other members of the *Trichophoromyia* may transmit *L.lainsoni*.

Transmission level: enzootic & zoonotic terrestrial.

L. (V.) shawi Lainson, Braga, Ishikawa & Silveira 1989

Host(s): monkeys - *Cebus apella* (BR), *Chiropotes satanas* (BR); sloths - *Choloepus didactylus*(BR), *Bradypus tridactlyus*(BR); coatimundis - *Nasua nasua*(BR).

Vector(s): *Lu.whitmani*(BR)

Ecology: Distributed south of the Amazon river. The arboreal cycle involves a variety of large mammals that reflects the catholic feeding habits of the vector. It is debatable if any one of them can be considered as being more important than another as a reservoir. This parasite is very common in man in both the States of Pará and Maranhão, Brazil. In some localities of the latter State there is evidence of zoonotic peridomestic transmission.

Capuchin monkeys (*C.apella*) are common in these locations and could be the source of infection.

Transmission level: enzootic principally arboreal; zoonotic terrestrial.

L. (V.) naiffi Lainson & Shaw. 1989

Host(s): nine-banded armadillos *Dasypus novemcinctus* (BR)

Vector(s): *Ps.ayrozai* (BR), *Ps.paraensis* (BR), *Ps.squamiventris squamiventris* (BR)

Ecology: There is only one mammalian host of this parasite but various species of sand flies appear to act as vectors. So far this appears to be an Amazonian species but it is quite likely that it exists outside the region since 9-banded armadillos are common throughout Latin America, as are *Psychodopygus* sand flies such as *Ps.ayrozai*. None of the sand flies that have been found infected are associated with armadillo burrows which suggests that transmission occurs at night when the armadillos are foraging for food.

Transmission level: enzootic, zoonotic terrestrial.

PARASITES ORIGINALLY DESCRIBED AS *LEISHMANIA* THAT DO NOT FALL INTO EITHER OF THE TWO NAMED SUBGENERA.

The hosts of the first three parasites, guinea pigs and porcupines, belong to rodent families (Caviidae, Erethizontidae) that are exclusively neotropical and unrelated to the families that are the principal hosts of the subgenera *L.* (*Leishmania*) and *L.* (*Viannia*). The other two are found in sloths that are exclusive to the neotropics. They appear to represent a separate evolutionary group linked (53, 54) to parasites of the genus *Endotrypanum* that occur in sloths (Xenathra). They have biological and molecular characters in common with the subgenus *L.*(*Viannia*) and some have a peripylarian development in the sand fly gut.

L. enriettii Muniz & Medina, 1948

Host(s): the domestic guinea pig (*Cavia porcellus* (BR - Paraná)). This parasite has only been found on three occasions including its discovery in 1945. Wild guinea pigs (*Cavia aperea*) do not develop lesions when inoculated with *L. enriettii* and some workers consider that this suggests that they are not the natural hosts. However, this is not a good argument since most leishmanial infections in wild animals are occult and are extremely difficult to detect.

Vector: The suspected vector is *Lu. monticola*. It was found in forest, near to the houses of the positive guinea pigs and has been experimentally infected.

Ecology: So far all infections have occurred in guinea pigs from areas that are close to pine (*Araucaria angustifolia*) forests in Paraná State, Brazil. There are sand flies in these forests and one, *Lu. monticola*, is susceptible to infection. A working hypothesis is that sand flies, that became infected when feeding on wild animal(s), accidentally transmitted the infection to domestic guinea pigs. The natural reservoir of this parasite, therefore, remains unknown.

Transmission level: unknown

L. hertigi Herrer, 1971

Host(s): Porcupines - *Coendou bicolor* (PA), *Sphiggurus mexicanus* (CR) In the original papers the hosts were listed as *Coendou rothschildi* (PA) that is considered as a synonym of *C.bicolor* and *Coendou mexicana* (CR) that is now considered as a species of the genus *Sphiggurus.*

Vector: unknown

Ecology: Its hosts are associated with Central American forests and the incidence is relatively high. There are no clues that might lead to the vector. Porcupines are nocturnal animals and excellent climbers but they also visit the ground and occasionally invade plantations, such as banana and corn, for food. Porcupines often sleep in hollow trees that are a known resting site of sand flies but so far no link has been found with such species.

Transmission level: unknown

L. deanei Lainson & Shaw, 1977

Host(s): Porcupines - *Coendou prehensilis*(BR), *Coendou* sp. (BR)

Vector(s): unknown

Ecology: This is a common parasite of porcupines of Amazonia and transmission must be efficient. However, so far no vector has been found and experimental infection in laboratory bred sand flies have failed. This parasite does not infect other species of mammals including laboratory rodents.

Transmission level: unknown

L. equatorensis Grimaldi, Kreutzer, Hashiguchi, Gomez, Mimory & Tesh, 1992

Hosts: the two-toed sloth *Choloepus hoffmanni*(EC), the squirrel *Sciurus granatensis* (EC).

Vector(s):unknown

Ecology: This appears to be a parasite transmitted amongst arboreal animals in climax-forest. It has only been described from Ecuador but it may occur in neighbouring countries.

Transmission level: presumably arboreal

L. colombiensis Kreutzer *et al.,* 1991

Host(s): two-toed sloths *Choloepus hoffmanni*(PA), the dog *Canis familiaris*(VE)

Vector(s):*Lu.hartmanni*(CO),*Lu.gomezi*(PA), *Ps.panamensis*(PA)

Ecology: The finding of this parasite in 3 different countries indicates a widely distributed enzootic cycle involving sloths and possibly other arboreal animals. It is not clear if sloths are the principal reservoir. The two isolations from man and dogs were made from bone marrow biopsies. Both are presumed to be accidental. Given the benign nature of the infections others may go unnoticed.

Transmission level: enzootic primarily arboreal, zoonotic terrestrial

OCCULT FERAL LEISHMANIAL INFECTIONS AND HUMAN DISEASE.

Leishmaniasis is an occult infection in most animals with the exception of dogs. It is also uncommon to find people with no clinical symptoms who have lived all their lives in endemic regions even though there is immunological evidence of infection. Is this due to innate immunity or some form of acquired immunity? It could be both but the pioneering work of David Sacks' group throws some light on this enigmatic situation. They showed (55) that exposure to the bites of uninfected sand flies protected against future infection. Both animals and man living in endemic areas are continuously bitten by sand flies since the day they are born which theoretically could protect them. If this proves to be true then we might expect that the severity of clinical symptoms in either animals or man could partly depend on how much they are bitten by uninfected sand flies.

ACKNOWLEDGEMENTS.

The author is grateful to many scientists for useful discussion and comments and in particular to José Eduardo Tolezano and Sinval Brandão Filho for introducing him to the ecology of leishmaniasis outside the Amazon region. This work was in part supported by grants from the Fundação de Amparo à Pesquisa do Estado de São Paulo (FAPESP), Brazilian Conselho Nacional de Pesquisa (CNPq).

REFERENCES

1. Lainson, R.and Shaw, J.J. 1979. "The Role of Animals in the Epidemiology of South American Leishmaniasis." In *Biology of the Kinetoplastida* , vol 2.Lumsden WHR, Evans DA, eds. London, New York & San Francisco: Academic Press, 1-116.

2. Lainson, R. and Shaw, J.J. 1987. "Evolution, classification and geographical distribution." In *The Leishmaniases in Biology and Medicine: Volume I Biology and Epidemiology* , vol I.Peters W, Killick-Kendrick R, eds. London: Academic Press Inc, 1-120

3. Shaw, J.J. and Lainson, R. 1987. "Ecology and epidemiology: New World." In *The Leishmaniases in Biology and Medicine: Volume I Biology and Epidemiology* .Peters W, Killick-Kendrick R, eds. London: Academic Press Inc, 291-363.

4. Grimaldi, G., Jr., Tesh, R.B., and McMahon-Pratt, D. 1989. A review of the geographic distribution and epidemiology of leishmaniasis in the New World. Amer J Trop Med Hyg 41: 687-725.

5.Young, D.G. and Duncan, M.A. , 1994. Guide to the identification and geographic distribution of *Lutzomyia* sand flies in Mexico, the West Indies, Central and South America (Diptera: Psychodidae).: Associated Publishers, American Entomological Institute

6. Shaw, J.J. 1999. "The relationship of sand fly ecology to the transmission of leishmaniasis in South America with particular reference to Brazil." In *Contributions to the knowledge of Diptera* , vol 14.Burger J, ed. Gainsville, Florida: Associated Publishers, 503-517

7. Fernandes, A.P., Nelson, K., and Beverley, S.M. 1993. Evolution of nuclear ribosomal RNAs in kinetoplastid protozoa: perspectives on the age and origins of parasitism. Proc Natl Acad Sci U S A 90: 11608-11612.

8. Shaw, J.J. 1997. Ecological and evolutionary pressures on leishmanial parasites. Braz J Gen 20: 123-128.

9. Thomaz-Soccol, V., Lanotte, G., Rioux, J.A., Pratlong, F., Martini-Dumas, A., and and Serres, E. 1993. Monophyletic origin of the genus *Leishmania* Ross, 1903. Ann Parasitol Hum Comp 68: 107-108.

10. Thomaz-Soccol, V., Lanotte, G., Rioux, J.A., Pratlong, F., Martini-Dumas, A., and Serres, E. 1993. Phylogenetic taxonomy of New World *Leishmania*. Ann Parasitol Hum Comp 68 : 104-106.

11. Noyes, H.A., Morrison, D.A., Chance, M.L., and Ellis, J.T. 2000. Evidence for a neotropical origin of *Leishmania*. Mem Instituto Oswaldo Cruz 95 : 575-578.

12. Tibayrenc, M., Kjellberg, F., and Ayala, F.J. 1990. A clonal theory of parasitic protozoa: the population structure of *Entamoeba* , *Giardia* , *Leishmania* , *Plasmodium* , *Trichomonas* and *Trypanosoma* and their medical and taxonomical consequences. Proc Nat Acad Sci USA 87: 2414-2418.

13.Sampaio, L.F. 1951. O aparecimento, a expansão e o fim da leishmaniose no Estado de São Paulo. Revta Bras Med 8: 717-721.

14. Peres, C.A. 1999. Ground fires as agents of mortality in a Central Amazonian forest. J. Trop. Ecol. 15: 535-541.

15. Silva, E.S., Gontijo, C.M., Pacheco, R.S., Fiuza, V.O., and Brazil, R.P. 2001. Visceral leishmaniasis in the Metropolitan Region of Belo Horizonte, State of Minas Gerais, Brazil. Mem Inst Oswaldo Cruz 96: 285-291.

16. Bush, M.B., Miller, M.C., De Oliveira, P.E., and Colinvaux, P.A. 2000. Two histories of environmental change and human disturbance in eastern lowland Amazonia. Holocene 10: 543-553.

17. Shaw, J.J. 1987. "Appendix I codes for hosts of *Leishmania* ." In *The Leishmaniases in Biology and Medicine: Volume I Biology and Epidemiology* .Peters W, Killick-Kendrick R, eds. London: Academic Press Inc, 465-472

18. Martinez, E., Le Pont, F., Mollinedo, S., and Cupolillo, E. 2001. A first case of cutaneous leishmaniasis due to *Leishmania* (*Viannia*) *lainsoni* in Bolivia. Trans R Soc Trop Med Hyg 95: 375-377.

19. Kreutzer, R.D., Corredor, A., Grimaldi, G., Jr., Grogl, M., Rowton, E.D., Young, D.G., Morales, A., McMahon-Pratt, D., et al. 1991. Characterization of *Leishmania colombiensis* sp. n (Kinetoplastida: Trypanosomatidae), a new parasite infecting humans, animals, and phlebotomine sand flies in Colombia and Panama. Amer J Trop Med Hyg 44: 662-675.

20. Katakura, K., Matsumoto, Y., Gomez, E.A., Furuya, M., and Hashiguchi, Y. 1993. Molecular karyotype characterization of *Leishmania panamensis* , *Leishmania mexicana* and *Leishmania major* -like parasites: agents of cutaneous leishmaniasis in Ecuador. Amer J Trop Med Hyg 48: 707-715.

21. Raccurt, C.P., Pratlong, F., Moreau, B., Pradinaud, R., and Dedet, J.P. 1995. French Guiana must be recognized as an endemic area of *Leishmania (Viannia) braziliensis* in South America. Trans R Soc Trop Med Hyg 89: 372.

22. Darie, H., Deniau, M., Pratlong, F., Lanotte, G., Talarmin, A., Millet, P., Houin, R., and Dedet, J.P. Cutaneous leishmaniasis of humans due to *Leishmania (Viannia) naiffi* outside Brazil. Trans R Soc Trop Med Hyg 89: 476-477.

23. Darce, M., Moran, J., Palacios, X., Belli, A., Gomez-Urcuyo, F., Zamora, D., Valle, S., Gantier, J.C., et al. 1995. Etiology of human cutaneous leishmaniasis in Nicaragua. Trans R Soc Trop Med Hyg 1991; 85: 58-59.

24. Yamasaki, H., Agatsuma, T., Pavon, B., Moran, M., Furuya, M., and Aoki, T. 1994. *Leishmania major* -like parasite, a pathogenic agent of cutaneous leishmaniasis in Paraguay. Amer J Trop Med Hyg 51: 749-757.

25. Franke, E.D., Lucas, C.M., Tovar, A.A., Kruger, J.H., De Rivera, M.V., and Wignall, F.S. 1990. Diffuse cutaneous leishmaniasis acquired in Peru. Amer J Trop Med Hyg 43: 260-262.

26. Lucas, C.M., Franke, E.D., Cachay, M.I., Tejada, A., Carrizales, D., and Kreutzer, R.D. 1994. *Leishmania (Viannia) lainsoni* : first isolation in Peru. Amer J Trop Med Hyg 51: 533-537.

27. Dujardin, J.C., Bañuls, A.L., Llanos-Cuentas, A., Alvarez, E., DeDoncker, S., Jacquet, D., Le Ray, D., Arevalo, J., et al. 1995. Putative *Leishmania* hybrids in the Eastern Andean valley of Huanuco, Peru. Acta Trop 59: 293-307.

28. Bonfante-Garrido, R., Meléndez, E., Barroeta, S., de Alejos, M.A., Momen, H., Cupolillo, E., McMahon-Pratt, D., and Grimaldi, G. 1992. Cutaneous leishmaniasis in western Venezuela caused by infection with *Leishmania venezuelensis* and *L . braziliensis* variants. Trans R Soc Trop Med Hyg 86: 141-148.

29. Monroy-Ostria, A., Hernandez-Montes, O., and Barker, D.C. 2000. Aetiology of visceral leishmaniasis in Mexico. Acta Trop 75: 155-161.

30. Christensen, H.A., de Vasquez, A.M., and Petersen, J.L. 1999. Short report epidemiologic studies on cutaneous leishmaniasis in eastern Panama. Amer J Trop Med Hyg 60: 54-57.

31. Ponce, C., Ponce, E., Morrison, A., Cruz, A., Kreutzer, R., McMahon-Pratt, D., and Neva, F. 1991. *Leishmania donovani chagasi* : new clinical variant of cutaneous leishmaniasis in Honduras. Lancet 337: 67-70.

32. Uliana, S.R.B., Ishikawa, I., Stempliuk, V.A., de Souza, A., Shaw, J.J., and Floeter-Winter, L.M. 2000. Geographical distribution of neotropical *Leishmania* analysed by ribosomal oligonucleotide probes. Trans R Soc Trop Med Hyg 94: 261-264.

33. Corredor, A., Gallego, J.F., Tesh, R.B., Pel ez, D., Diaz, A., Montilla, M., and Pal u, M.T. 1989. *Didelphis marsupialis* , an apparent wild reservoir of *Leishmania donovani chagasi* in Colombia, South America. Trans R Soc Trop Med Hyg 83: 195.

34. Zulueta, A.M., Villarroel, E., Rodriguez, N., Feliciangeli, M.D., Mazzarri, M., Reyes, O., Rodriguez, V., Centeno, M., et al. 1999. Epidemiologic aspects of American visceral leishmaniasis in an endemic focus in Eastern Venezuela. Amer J Trop Med Hyg 61: 945-950.

35. Courtenay, O., Santana, E.W., Johnson, P.J., Vasconcelos, I.A., and Vasconcelos, A.W. 1996. Visceral leishmaniasis in the hoary zorro *Dusicyon vetulus* : a case of mistaken identity. Trans R Soc Trop Med Hyg 90: 498-502.

36. Ward, R.D., Phillips, A., Burnet, A., and Marcondes, C.R. , 1988. "The *Lutzomyia longipalpis* complex: reproduction and distribution." In *Biosystematics of haematophagus insects* , vol The Sytematics Association Special Publication 37.Service MW, ed. Oxford: Clarendon Press, 257-269.

37. Lanzaro, G.C., Warburg, A., Lawyer, P.G., Herrero, M.V., and Ostrovska, K. 1993. *Lutzomyia longipalpis* is a species complex: genetic divergence and interspecific hybrid sterility among three populations. Amer J Trop Med Hyg; 48: 839-847.

38. Travi, B.L., Velez, I.D., Brutus, L., Segura, I., Jaramillo, C., and Montoya, J. 1990. *Lutzomyia evansi* , an alternate vector of *Leishmania chagasi* in a Colombian focus of visceral leishmaniasis. Trans R Soc Trop Med Hyg 84: 676-677.

39. dos Santos, S.O., Arias, J., Ribeiro, A.A., de Paiva Hoffmann, M., de Freitas, R.A., and Malacco, M.A. 1998. Incrimination of *Lutzomyia cruzi* as a vector of American visceral leishmaniasis. Med Vet Entomol 12: 315-317.

40. Chable-Santos, J.B., Van Wynsberghe, N.R., Canto-Lara, S.B., and Andrade-Narvaez, F.J. 1995. Isolation of *Leishmania* (*L* .) *mexicana* from wild rodents and their possible role in the transmission of localized cutaneous leishmaniasis in the state of Campeche, Mexico. Amer J Trop Med Hyg 53: 141-145.

41. Hashiguchi, Y., Gomez, E.A., de Coronel, V.V., Mimori, T., Kawabata, M., Furuya, M., Nonaka, S., Takaoka, H., et al. 1991. Andean leishmaniasis in Ecuador caused by infection with *Leishmania mexicana* and *L* . *major* -like parasites. Amer J Trop Med Hyg 44: 205-217.

42. Van Wynsberghe, N.R., Canto-Lara, S.B., Damian-Centeno, A.G., Itza-Ortiz, M.F., and Andrade-Narvaez, F.J. 2000. Retention of *Leishmania* (*Leishmania*) *mexicana* in naturally infected rodents from the State of Campeche, Mexico. Mem Inst Oswaldo Cruz 95: 595-600.

43. Mimori, T., Grimaldi, G., Jr., Kreutzer, R.D., Gomez, E.A., McMahon-Pratt, D., Tesh, R.B., and Hashiguchi, Y. 1989. Identification, using isoenzyme electrophoresis and monoclonal antibodies, of Leishmania isolated from humans and wild animals of Ecuador. Amer J Trop Med Hyg 40: 154-158.

44. Martinez, E., Le Pont, F., Torrez, M., Telleria, J., Vargas, F., Dujardin, J.C., and Dujardin, J.P. 1999. Lutzomyia nuneztovari anglesi (Le pont & Desjeux, 1984) as a vector of Leishmania amazonensis in a sub-Andean leishmaniasis focus of Bolivia. Amer J Trop Med Hyg 61: 846-849.

45. Brandão-Filho, S.P., Brito, M.E.F., Carvalho, F.G., Cupolillo, E., and Shaw, J.J. 2001. *Bolomys lasiurus* and other silvatic and sinanthropus mammals as reservoirs of *Leishmania* (*V* .) *braziliensis* in Atlantic Rain Forest, region of Pernambuco, Brazil. J Bras Patol 37: 35.

46. Alexander, B., Lozano, C., Barker, D.C., McCann, S.H., and Adler, G.H. 1998. Detection of *Leishmania (Viannia) braziliensis* complex in wild mammals from Colombian coffee plantations by PCR and DNA hybridization. Acta Trop 69: 41-50.

47. Feliciangeli, M.D., Rodriguez, N., Bravo, A., Arias, F., and Guzman, B. 1994. Vectors of cutaneous leishmaniasis in north-central Venezuela. Med Vet Entomol 8: 317-324.

48. Rowton, E.D., de Mata, M., Rizzo, N., Porter, C.H., and Navin, T.R. 1992. Isolation of *Leishmania braziliensis* from *Lutzomyia ovallesi* (Diptera:Psychodidae) in Guatemala. Amer J Trop Med Hyg 46: 465-468.

49. da Silva, O.S. and Grunewald, J. 1999. Contribution to the sand fly fauna (Diptera: Phlebotominae) of Rio Grande do Sul, Brazil and *Leishmania* (*Viannia*) infections. Mem Inst Oswaldo Cruz 94: 579-582.

50. Cupolillo, E., Grimaldi Junior, G., Momen, H., and Beverley, S.M. 1995. Intergenic region typing (IRT): a rapid molecular approach to the characterization and evolution of *Leishmania* . Mol Biochem Parasitol 73: 145-155.

51. Ishikawa, E.A.Y., Silveira, F.T., Magalhães, A.L.P., Guerra Jnr, R.B., Melo, M.N., Gomes, R., Silveira, T.G.V., and Shaw, J.J. Genetic variation in populations of *Leishmania* species in Brazil . Trans R Soc Trop Med Hyg 2002; (in press).

52. Llanos-Cuentas, E.A., Roncal, N., Villaseca, P., Paz, L., Ogusuku, E., Perez, J.E., Caceres, A., and Davies, C.R. 1999. Natural infections of *Leishmania peruviana* in animals in the Peruvian Andes. Trans R Soc Trop Med Hyg 93: 15-20.

53. Croan, D.G., Morrison, D.A., and Ellis, J.T. 1997. Evolution of the genus *Leishmania* revealed by comparison of DNA and RNA polymerase gene sequences. Mol Biochem Parasitol 89: 149-159.

54. Cupolillo, E., Pereira, L.O., Fernandes, O., Catanho, M.P., Pereira, J.C., Medina-Acosta, E., and Grimaldi, G., Jr. 1998. Genetic data showing evolutionary links between *Leishmania* and *Endotrypanum* . Mem Inst Oswaldo Cruz 93: 677-683.

55. Kamhawi, S., Belkaid, Y., Modi, G., Rowton, E., and Sacks, D. 2000 Protection against cutaneous leishmaniasis resulting from bites of uninfected sand flies. Science; 290: 1351-1354.

PHLEBOTOMINE SAND FLIES: BIOLOGY AND CONTROL

Robert Killick-Kendrick

Biology Department, Imperial College at Silwood Park, Ascot, Berks SL5 7PY, U.K.

INTRODUCTION

Of approximately 800 species or subspecies of phlebotomine sand flies, 80 are proven or probable vectors of the 22 species of *Leishmania* that cause human disease (Tables 1 and 2). In some foci of leishmaniasis, the vectors are unknown and it is certain that more species will be added to the list. With the exceptions of *Lutzomyia longipalpis* of the New World and *Phlebotomus papatasi, P. ariasi* and *P. perniciosus* of the Old World, the biology of most of these vectors is poorly known. This is a constraint in devising means of controlling the vectors and thus reducing the risk of infection. Most importantly, so little is known about the terrestrial breeding sites of sand flies that any attempt to control them by attacking the peri-marginal stages is impractical.

In this chapter a summary is given of the biology of sand flies and the options for their control are reviewed. A short guide to methods for the study of sand flies is included with key references to the literature.

TAXONOMY AND SYSTEMATICS

Sand flies are grouped in the Suborder Nematocera of the Order Diptera (two-winged flies). Below that, the classification is not universally agreed, but they are generally put in the Family Psychodidae, subfamily Phlebotominae. Lane (1) recognises six genera of which only two are of medical importance, namely *Phlebotomus* of the Old World, divided into 12 subgenera, and *Lutzomyia* of the New World, divided into 25 subgenera and species groups. All proven vectors of the leishmaniases are species of these two genera (Tables 1 and 2). A key reference to the taxonomy of Old World sand flies is Seccombe *et al.*(2); for taxa of the genus *Lutzomyia* of the New World, see Young and Duncan (3) . Modern approaches to the taxonomy and systematics of sand flies include isoenzyme analyses(4,5,6), numerical taxonomy (7,8), multivariate analyses of morphological characters (9,10,11,12), DNA sequencing (13,14,15) and the preparation of CD-ROMs with illustrated non-dichotomous keys (16).

causative organism	proven or probable vector	places of known or suspected transmission by listed vector
Le donovani s.l.	*P. orientalis*†	Ethiopia; Saudi Arabia; Yemen; Sudan
	P. argentipes†	India; Bangladesh; Nepal
	P. martini†	Kenya; Ethiopia
	P. celiae†	Kenya; Ethiopia
	P. alexandri†	China
Le. infantum	*P. ariasi*†	W. Mediterranean
	P. kandelakii	Transcaucasus; Iran; Afghanistan
	P. langeroni†	Egypt
	P. longicuspis	N. Africa
	P. neglectus†	E. Mediterranean
	P. perfiliewi†	Italy; E. Mediterranean; N. Africa
	P. perniciosus†	W. and C. Mediterranean
	P. smirnovi	China; Kazakhstan
	P. tobbi	E. Mediterranean; Sicily
	P. transcaucasicus	Azerbaijan
	P chinensis	China
	P. longiductus	Central Asia
Le tropica	*P. sergenti*†	N. Africa; Middle East; Afghanistan; Iran; Transcaucasus; E.Mediterranean
Le. tropica	*P. guggisbergi*†	Kenya
	P. aculeatus	Kenya
Leishmania sp.	near *P. rossi*§	Namibia
Le major	*P. duboscqi*†	W. Africa; Kenya; Ethiopia
Le major	*P. papatasi*†	N. Africa; Sudan; Central Asia; Near and Middle East; Indian Subcontinent
	P. salehi	NW India; Iran
	P. caucasicus	Iran; Central Asia
	P. alexandri	Turkmenistan
	P. ansarii	Iran
Le. aethiopica	*P. longipes*†	Ethiopia
	P. pedifer†	Kenya; Ethiopia

Table 1. *Proven (†) and probable vectors of the Old World leishmaniases.*
See Killick-Kendrick (40,41) for references. (See Davidson (11) for the differences between this sand fly and P. rossi.)

causative organism	proven or probable vector	Countries of known or suspected transmission by listed vector
Le.infantum *	*Lu. longipalpis†*	Argentina; Bolivia; Brazil; Colombia; Costa Rica; El Salvador; Guatamala; Guyana;Honduras; Mexico; Nicaragua; Paraguay; Suriname;
	Lu. evansi	Colombia; Costa Rica; Venezuela
Le. garnhami‡	*Lu. youngi*	Costa Rica; Venezuela
Le. peruviana	*Lu. peruensis*	Peru
	Lu. verrucarum	Peru
Le. lainsoni	*Lu. ubiquitalis†*	Brazil; Peru
Le. shawi	*Lu. whitmani†*	Brazil
Le. naiffi	*Lu. squamiventris*	Brazil
Le. colombiensis	*Lu. hartmanni†*	Colombia
	Lu. gomezi	Panama; Venezuela
	Lu. panamensis	Panama; Venezuela
Le. guyanensis	*Lu. umbratilis†*	Brazil; French Guiana; Colombia
	Lu. anduzei†	Brazil; French Guiana
Le. braziliensis s.l.	*Lu. wellcomei†*	Brazil
	Lu. complexus†	Brazil
	Lu. intermedia	Brazil
	Lu. pessoai	Brazil
	Lu. migonei	Brazil
	Lu. amazonensis	Brazil
	Lu. paraensis	Brazil
	Lu. whitmani†	Brazil
	Lu. panamensis	Venezuela
	Lu. ovallesi†	Venezuela
	Lu. yucumensis	Bolivia
	Lu. llanosmartinsi	Bolivia
	Lu. c. carrerai†	Bolivia
	Lu. ayrozai	Bolivia
	Lu. spinicrassa	Colombia
	Lu colombiana[151]	Colombia
	Lu. pia	Colombia
	Lu. townsendi	Colombia
Le. panamensis	*Lu. trapidoi†*	Panama; Colombia
	Lu. gomezi	Panama
	Lu. panamensis	Panama
	Lu. ylephiletor	Panama
Le. mexicana	*Lu. o. olmeca†*	Belize; Costa Rica; Guatamala; Honduras; Mexico
	Lu. ayacuchensis†	Ecuador
	Lu. ylephiletor	Guatamala
Le. amazonensis	*Lu. flaviscutellata†*	Bolivia; Brazil; Colombia; Ecuador; French Guiana; Paraguay; Venezuela
	Lu. olmeca nociva	Brazil
	Lu. reducta	Brazil; Venezuela
Le. venezuelensis	*Lu. olmeca bicolor*	Venezuala
Leishmania sp.	*Lu. anthophora*	USA; Mexico
Leishmania sp.	*Lu. christophei*	Dominican Republic
Le. pifanoi‡	*Lu. flaviscutellata*	Venezuela

Table 2. *Proven (†) and probable vectors of the New World leishmanias See Killick-Kendrick(40,41) for references. *Some workers retain the old name Le.chagasi for New World Le. infantum. Mauricio etal.(58) present strong evidence for the common identity of these two parasites. Some workers do not currently accept the identity of this species.*

BIOLOGY OF SAND FLIES
Distribution
The distribution of phlebotomine sand flies extends northwards to just above latitude 50°N in south west Canada and just below this latitude in northern France and Mongolia. Their southernmost distribution is at about latitude 40°S, but they are absent from New Zealand and the Pacific islands. Their altitudinal distribution is from below sea level (by the Dead Sea)[36] to 3300 m above sea level in Afghanistan (*P. rupester*) (17).

Life cycle
Female sand flies take blood to mature their eggs. Males do not bite. Females of some populations are autogenous and lay their first batch of eggs without having taken a blood meal. One species (*Lu. mamedei*) is both autogenous and parthenogenetic (18). The time from a blood meal to maturation of eggs ranges from about 4-8 days. Soil bacteria attract gravid *P. papatasi* (19) and conspecific eggs act as an oviposition attractant or stimulant. The active principle has been isolated from eggs of *Lu. Longipalpis* (20) and shown to be produced by the accessory glands of the female (21). Gravid females of *Lu. longipalpis* are also attracted by hexanol and 2-methyl-2-butanol in chicken or rabbit faeces (22). When used together, these two chemicals elicit the same oviposition responses as total volatile extracts. On the available evidence from studies on only two species, it seems probable that gravid sand flies are responding to combined effects.

Preimaginal stages are terrestrial, not aquatic. There are four larval instars. Palaearctic species diapause in winter as 4th instar larvae and some tropical species survive heavy rains by diapausing as eggs. Larval breeding sites are rich in organic matter (23) and must remain humid throughout the development of the larvae. In the laboratory, the time of larval development from egg to pupa is four weeks or more depending upon the species and temperature. The development of the sessile pupae takes about 10 days.

A few species mate on the host where the males lek before the females arrive to take a blood meal (24,25). Mate recognition is by pheromones of males (26), songs (27) or, for one species, males riding on the back of the females (28). Mating may occur before, during or after blood meals and may be more than once in a single gonotrophic cycle (29,30). The choice of host by the females varies with species. Some feed on only one host (e.g. bats: *Lu. vespertilionis*), others may feed on a very limited range (e.g. bovids and man: *P. argentipes*), or on any mammal (e.g. *P. ariasi*), or any mammal, bird or reptile (e.g. *P. papatasi*). As with other Nematocera, adult phlebotomines of both sexes feed on natural sources of sugar which, for all species studied, is honeydew of aphids (31) or, perhaps, coccids. One species (*P. papatasi*) also feeds on plant sap (32).

As with all other haematophagous arthropods, blood feeding by sand flies is preceded by the salivation into the skin of the hosts. Saliva of sand flies contains many pharmacologically active molecules including anticoagulants and vasodilators. The saliva is currently receiving much attention mainly because of experimental evidence that prior exposure to sand fly saliva moderates the subsequent course of a leishmanial infection in mice (33).

Activity

Diurnal resting places of adult phlebotomines include houses; latrines; cellars; stables; caves; fissures in walls, rocks or soil; dense vegetation; tree holes and buttresses; burrows of rodents and other mammals; bird's nests; and termitaria. The activity of adult sand flies is crepuscular and nocturnal, although some Neotropical forest species will also bite during the day. Peak biting is commonly immediately after sunset. Human kairomones and CO_2 both play a part in attraction to man, with the former more important than the latter (34). The maximum distance of dispersion recorded in mark-release-recapture field experiments is 2.3 km (35). Flight speed is about 1m/sec (36).

Vectorial roles

Only a minority of sand flies are vectors of leishmaniasis. Moreover, many species that are proven vectors of one species of *Leishmania* cannot support the growth of other species. This is thought to be because they lack appropriate lipophosphoglycan ligands (37) or, perhaps, lectin binding sites (38) on the internal surface of the midgut to which the parasites must attach. With some vector species, there is pattern of susceptibility to *Leishmania* suggesting that parasites and sand flies evolved together resulting in closely related sand flies, assumed to have had a common ancestor, acting as vectors of the same species of *Leishmania* (39).

The principal generally accepted criteria for the incrimination of a vector are evidence that it is anthropophilic and the isolation and typing of parasites from wild-caught sand flies that are indistiguishable from strains isolated from patients. Supporting experimental evidence is that the sand fly can support the development of the parasite after the infecting blood meal has been digested and voided, and is able to transmit the parasite by bite (40).

CONTROL OF SAND FLIES

There are no vaccines or prophylactic drugs for the leishmaniases and a reduction of the risk of infection to man depends upon attacking the vectors, the reservoir hosts or both. References to the literature on the control of sand flies are in a recent review by Killick-Kendrick (41). There are four main approaches none of which is likely to succeed without stability, political commitment, economic means, public education and trained staff.

The complete destruction of the habitat followed by exploitation of the land by settlements, farming or forestry is the only permanent way of controlling sand flies. It was used with success but at great expense by Soviet workers in the highly endemic foci of zoonotic cutaneous leishmaniasis (*L. major* transmitted by *P. papatasi*) in the central Asian Republics of the former Soviet Union.

House spraying with insecticides to control malaria in the 1950s and 60s was accompanied by a fall in the incidence of kala azar (visceral leishmaniasis) in India (*L. donovani* transmitted by *P. argentipes*). Similar falls were recorded in several other countries. This gives good results only with the few species of sand flies that are strongly endophilic and, because of the high cost and need to spray for an indefinite period, WHO recommends that this means of control should be applied only during epidemics of serious forms of leishmaniasis likely to lead to high death rates. Insecticide fogging of settlements is an alternative to house spraying, although concerns for the environment make this an unattractive option.

Protection of the population with insecticide impregnated bednets (42) or curtains (43) as used in the control of malaria may reduce the risk of forms of leishmaniasis that are transmitted peridomestically e.g. anthroponotic cutaneous leishmaniasis of the Old World (*L. tropica* transmitted by *P. sergenti*). Success depends upon the behaviour of the target species of sand fly, public education, a well organised primary health care service and the motivation, behaviour and economic means of the community.

In foci where dogs are reservoir hosts of *L. infantum*, the elimination of all dogs, as in eastern China, is an efficient means of control. This would not be acceptable in other foci. The selected culling of seropositve dogs as investigated in Brazil was followed by no more than a temporary fall in the prevalence of canine infections (44) and is expensive (45). A cheaper and better way may be by fitting dogs with deltamethrin-impregnated collars (Scalibor® ProtectorBands, Intervet International) which act as a long term depot slowly releasing the insecticide into the lipids of the skin. In a comparison of four topically applied insecticides, it was concluded that deltamethrin-impregnated collars are the one of choice (46) and the results of several studies suggest the collars may be a promising new public health measure for the widespread control of canine leishmaniasis leading to a reduction in the accompanying risk of infection to the human population. In laboratory experiments with three species of phlebotomines (*P. perniciosus* (47), *Lu. Longipalpis* (48) and *Lu. Migonei* (48)), deltamethrin impregnated collars have been shown to protect dogs from 80-96% of sand fly bites for seven to eight months. In Iran, guard dogs were protected against 80% of bites of *P. papatasi* eight days after collars were attached (49). No tests were done after this time. In the first field trial Maroli and colleagues (50) reported that, when the force of transmission in Italy was judged to be high, the estimated level at which dogs were protected by the collars from infection with *L. infantum* was 86%. There was no measurable difference in seroconversion between collared and uncollared dogs when the force of infection was low. Trials are in progress in foci of human visceral leishmaniasis with a canine reservoir to see if the expected fall in the prevalence of infection in dogs provided by the collars will be accompanied by a reduction in risk of visceral leishmaniasis to the human population. In some foci of leishmaniasis, the expectation of life of dogs is short, especially when they are not vaccinated against distemper or hardpad. This results in a rapid turnover of dogs requiring constant surveillance to ensure new additions to the canine population are collared and lost or stolen collars are replaced. As with all interventions, public education will be essential so that the control measure is understood and accepted by the community. Compared to house spraying, collars are cheaper, and expensive equipment and highly trained staff are not required for attachment and surveillance.

TECHNIQUES FOR THE STUDY OF SAND FLIES
Sampling.
Sampling methods are reviewed by Killick-Kendrick (51), Young and Duncan (3) and Alexander (52). Live sand flies can be collected by several different techniques. They can be caught (a) by an active search with sucking tubes and a torch in their resting places by day or night, (b) coming to bite man or livestock, (c) with a Shannon trap or (d) with CDC miniature light traps. Man landing catches are not only generally productive but also indicate species that are anthropophilic. In comparatively dry habitats in the Old World, sand fly surveys are often done with sticky paper traps. These are standard sized sheets of paper covered in castor oil. Engine oil can be used as an alternative to castor oil.

Preservation and mounting on slides
Sandflies are not pinned but are mounted on slides under coverslips in either Canada balsam or Berlese's fluid. Methods for preservation and mounting are given by (a)Young and Duncan (3) and (b) Lane (1). The flies can be conserved as dry specimens or, more usually, are stored in 70% ethanol. Long exposure to the alcohol makes the flies hard and difficult to mount. This can be avoided by transferring them from ethanol and storing them in Berlese's fluid.

Rearing and Colonisation

A bibliography of the colonisation of sand flies (53) gives references to all methods published up to 1991. Techniques for the initial establishment of colonies are given by Killick-Kendrick, M. and Killick-Kendrick, R. (54). For recent advances in colony maintenance, see (a) Killick-Kendrick, R. and Killick-Kendrick, M.(55) and (b) Killick-Kendrick, M. *et al* (56).

The terrestrial preimaginal stages are reared in Neoprene pots. Most workers feed the larvae on a finely ground composted mixture of equal parts rabbit faeces and commercially available rabbit chow. (57) The pots are kept humid by standing them on damp sand or filter paper in a snap top plastic food container. The optimal rearing temperature varies with species from 24-28°C. Adult sand flies must be provided with sugar which is usually a strong aqueous solution of sucrose on cotton wool. Blood meals are from hamsters, rabbits, chicken or a human volunteer. Development is comparatively slow taking at least six weeks from one blood meal to another.

Field studies

Techniques for (a) mark-release-recapture of sand flies, (b) studies on flight speed with a wind tunnel, (c) blood meal identification, (d) larval extraction from soil samples and (e) the dissection of sand flies for the detection and isolation of *Leishmania* are given by Killick-Kendrick (50).

REFERENCES

1.Lane, R.P. 1993. Sandflies (Phlebotominae). In: Medical Insects and Arachnids, (Eds Lane, R.P. and Crosskey, R.W.), London: Chapman and Hall 78-119.

2.Seccombe, A.K., Ready, P.D., and Huddleston, L.M. 1993. A Catalogue of Old World phlebotomine sandflies (Diptera: Psychodidae). Occasional Papers on Systematic Entomology 8: 1-57.

3.Young, D.G. and Duncan, M.A. 1994. Guide to the identification and geographic description of Lutzomyia sand flies in Mexico, the West Indies, Central and South America, Gainsville: Associated Publishers 1-881.

4.Dujardin, J.P., Le Pont, F. and Martinez, E. 2001. Quantitatve phenetics and taxonomy of some Phlebotomine taxa. Memorias do Instituto Oswaldo Cruz 94: 735-741.

5.Remy-Christensen, A., Perrotey, S., Pesson, B., Garcia-Stoeckel, M., Ferté, H., Morillas-Marquez, F. and Léger, N. 1986. Phlebotomus sergenti Parrot, 1917: morphological and isoenzymic comparisons of two natural populations from Tenerife (Canary Islands, Spain) and Crete (Greece). Parasitology Research 82: 48-51.

6.Voltz-Kristensen, A., Pesson, B., Léger, N., Madulo-Leblond, G., Killick-Kendrick, R. and Killick-Kendrick, M. 1991. Phosphoglucomutase in phlebotomine sandflies of the subgenus Larroussius from Corfu island, Greece. Medical and Veterinary Entomology 5: 135-137.

7.Rispail, P. and Léger, N. 1998. Numerical taxonomy of Old World Phlebotominae (Diptera: Psychodidae). Considerations of morphological characters in the genus Phlebotomus Rondani and Berté 1840. Memorias do Instituto Oswaldo Cruz 93: 773-785.

8.Rispail, P. and Léger, N. 1998. Numerical taxonomy of Old World Phlebotominae (Diptera: Psychodidae). Restatement of classification upon subgeneric morphological characters. Memorias do Instituto Oswaldo Cruz 93: 787-793.

9.Añez, N., Valenta, D.T., Cazorla, D., Quicke, D.J. and Feliciangeli, D.M. Multivariate analysis to discriminate species of phlebotomine sand flies (Diptera: Psychodidae): Lutzomyia townsendi, L. spinicrassa, and L. youngi. 1997. Journal of Medical Entomology 34: 312-316.

10.Cazorla, D., Añez, N., Marquez, V. and Nieves, E. 1992. Multivariate analysis to discriminate species of phlebotomine sandflies. (Diptera: Psychodidae). Lutzomyia cayennensis and Lutzomyia yencanensis. Revista de la Facultad de Medicina, Universidad de Los Andes 1: 120-124.

11.Davidson, I.H. 1993. Multivariate discrimination of Afrotropical females in the subgenus Synphlebotomus, putative vectors of leishmaniasis in subsaharan Africa (Diptera, Psychodidae, Phlebotominae). In: Entomologist extraordinary: festschrift in honour of Botha de Meillon, (Ed. Coetzee, M.) Johannesburg: South African Institute for Medical Research 18-29.

12.Lane, R.P. and Ready, P.D. 1985. Multivariate discrimination between Lutzomyia wellcomei, a vector of mucocutaneous leishmaniasis, and Lu. complexus (Diptera: Psychodidae). Annals of Tropical Medicine and Parasitology 79: 469-472.

13.Depaquit, J., Ferté, H., Léger, N., Killick-Kendrick, R., Rioux, J.-A., Killick-Kendrick, M., Hanafi, H.A. and Gobert, S. 2000. Molecular systematics of the Phlebotomine sandflies of the subgenus Paraphlebotomus (Diptera, Psychodidae, Phlebotomus) based on ITS 2 rDNA sequences. Hypotheses of dispersion and speciation. Insect Molecular Biology 9: 293-300.

14.Depaquit, J., Perrotey, S., Lecointre, G., Tillier, A., Tillier, S., Ferté, H., Kaltenbach, M. and Léger. N. 1998. Systematique moléculaire des Phlebotominae: étude pilote. Paraphylie du genre Phlebotomus. Compte rendu d'Académie des Sciences, Paris, Science de la Vie 321: 849-855.

15.Esseghir, S., Ready, P.D., Killick-Kendrick, R. and Ben Ismail, R. 1997. Mitochondrial haplotypes and phylogeography of Phlebotomus vectors of Leishmania major. Insect Molecular Biology 6: 211-225.

16.Niang, A.A., Geoffroy, G., Angel, G., Trouillet, J., Killick-Kendrick, R., Hervy, J.-P. and Brunhes, J. Les Phlébotomes d'Afrique Ouest. 2000. CD Rom, Paris, IRD Editions.

17.Artemiev, M.M. 1980. A revision of sandflies of the subgenus Adlerius (Diptera, Phlebotominae, Phlebotomus) (in Russian). Zoolohicheskiy Zhurnal 59: 1177-1192.

18.Brazil, R.P. and Oliveira, S.M.P. 1999. Parthenogenisis in the sandfly Lutzomyia mamedei. Medical and Veterinary Entomology 13: 463-464.

19.Radjame, K., Srinivasan, R. and Dhanda, V. 1997. Oviposition response of phlebotomid sandfly Phlebotomus papatasi to soil bacteria isolated from natural breeding habitats. Indian Journal of Experimental Biology 35: 59-61.

20.Dougherty, M.J., Hamilton, J.G.C. and Ward, R.D. 1994. Isolation of oviposition pheromone from eggs of the sandfly Lutzomyia longipalpis. Journal of Medical Entomology 31: 119-124.

21.Dougherty, M,J., Ward, R.D. and Hamilton, G. Evidence for the accessory glands as the site of production of the oviposition attractant and/or stimulant of Lutzomyia longipalpis (Diptera: Psychodidae). Journal of Chemical Ecolology, 18: 1165-1175.

22.Dougherty, M.J., Guerin, P.M. and Ward, R.D. 1995. Identification of oviposition attractants for the sandfly Lutzomyia longipalpis (Diptera: Psychodidae) in volatiles of faeces from vertebrates. Physiological Entomology 20: 23-32.

23.Bettini, S. and Melis, P. 1998. Leishmaniasis in Sardinia. 3. Soil analysis of a breeding site of three species of sandflies. Medical and Veterinary Entomology 2: 67-71.

24.Kelly, D.W. and Dye, C. 1997. Pheromones, kairomones and the aggregation dynamics of the sandfly Lutzomyia longipalpis. Animal Behaviour 53: 721-731.

25.Lane, R.P., Pile, M.M. and Amerasinghe, F.P. 1990. Anthropophagy and aggregation behaviour of the sandfly Phlebotomus argentipes in Sri Lanka. Medical and Veterinary Entomology 4: 79-88.

26.Nigam, Y. and Ward, R.D. 1991. Male sandfly pheromone and artificial host factors as attractants for female L. longipalpis. Physiological Entomology 16: 305-312.

27.De Souza, N.A., Ward, R.D., Hamilton, J.G.C., Kyriacou, C.P. and Peixoto, A. A. Copulation songs in three siblings of Lutzomyia longipalpis (Diptera: Psychodidae). Transactions of the Royal Society of Tropical Medicine, 96: in press.

28.Valenta, D., Killick-Kendrick, R. and Killick-Kendrick, M. 2000. Courtship and mating in Phlebotomus duboscqi (Diptera: Psychodidae), a vector of zoonotic cutaneous leishmaniasis in the Afrotropical Region. Medical and Veterinary Entomology 13: 207-210.

29.Guilvard, E., Rioux, J., Jarry, D. and Moreno, G. 1985. Accouplements successifs chez Phlebotomus ariasi Tonnoir, 1921. Annales de Parasitologie humaine et compare 60: 503.

30.Yuval, B. and Schlein, Y. 1986. Evidence for polygamy in Phlebotomus papatasi Scopoli. Annales de Parasitologie Humaine et Comparee 61: 693-694.

31.Killick-Kendrick, R. and Killick-Kendrick, M. 1987. Honeydew of aphids as a source of sugar for Phlebotomus ariasi. Medical and Veterinary Entomology 1: 297-302.

32.Schlein, Y. and Warburg A. 1986. Phytophagy and the feeding cycle of Phlebotomus papatasi (Diptera: Psychodidae) under experimental conditions. Journal of Medical Entomology 23: 11-15.

33.Valenzuela, J.C., Belkaid, Y., Garfield, M.K., Mendez, S., Kamhawi, S., Rowton, E.D., Sacks, D. and Ribeiro, J.M.C. 2001. Toward a defined anti-Leishmania vaccine targeting vector antigens: characterization of a protective salivary protein. Journal of Experimental Medicine 19: 331-342.

34.Pinto, M.C., Cambell-Lenrum, D.H., Lozovei, A.L., Teodoro, U. and Dacies, C.R. 2001. Phlebotomine sandfly responses to carbon dioxide and human odour in the field. Medical and Veterinary Entomology 15: 132-139.

35.Killick-Kendrick, R., Rioux, J-A., Bailly, M., Guy, M.W., Wilkes, T.J., Guy, F.M., Davidson, I., Knechtli, R., Ward, R.D., Guilvard, E., Périères, J. and Dubois, H. 1984. Ecology of leishmaniasis in the south of France. 20. Dispersal of Phlebotomus ariasi Tonnoir, 1921 as a factor in the spread of visceral leishmaniasis in the Cévennes. Annales de Parasitologie humaine et compare 59: 555-572.

36.Killick-Kendrick, R., Wilkes, T.J., Bailly, M., Bailly, I. and Righton, L.A. 1986. Preliminary field observations on the flight speed of a phlebotomine sandfly. Transactions of the Royal Society of Tropical Medicine and Hygiene 80: 138-142.

37.Sacks, D. and Kamhawi, S. 2001. Molecular aspects of parasite-vector and vector-host interactions in leishmaniasis. Annual Review of Entomology 55: 453-483.

38.Volf, P. and Killick-Kendrick, R. 1996. Post-engorgement dynamics of haemagglutination activity in the midguts of six species of phlebotomine sandflies. Medical and Veterinary Entomology 10: 247-250.

39.Killick-Kendrick, R. 1985. Some epidemiological consequences of the evolutionary fit between leishmaniae and their phlebotomine vectors. Bulletin de la Société de Pathologie Exotique 78: 747-755.

40.Killick-Kendrick, R. 1990. Phlebotomine vectors of the leishmaniases: a review. Medical and Veterinary Entomology 4: 1-24.

41.Killick-Kendrick, R. 1999. The biology and control of phebotomine sand flies. Clinics in Dermatology 17: 279-289.

42.Elnaiem, D.A., Elnahas, A.M. and Aboud, M.A. 1999. Protective efficacy of lamdacyhalothrin-impregnated bednets against Phlebotomus orientalis, the vector of visceral leishmaniasis in Sudan. Medical and Veterinary Entomology 13: 310 – 314.

43.Elnaiem, D.A., Aboud, M.A., El Mubarek, S.G., Hassan, H.K. and Ward, R.D. 1999. Impact of pyrethroid-impregnated curtains on Phlebotomus papatasi sandflies indoors at Khartoum. Medical and Veterinary Entomology 13: 191-197.

44.Ashford, D.A., David, J.R., Freire, M., David, R., Sherlock, I., Eulalio, M.C., Sampaio, D.P. and Badaro, R. 1998. Studies on the control of visceral leishmaniasis: impact of dog control on canine and human visceral leishmaniasis in Jacobina, Brazil. American Journal of Tropical Medicine and Hygiene 59: 53-57.

45.Akhaven, D. 1996. Análise de custo-efetividade de componte de leishmaniose no projeto de contrôle de doenças endêmica no nordeste do Brasil. Revista dePatologia Tropical 25: 203-252.

46.Reithinger, R., Teodoro, U. and Davies, C.R. 2001. Topical insecticide treatments to protect dogs from sand fly vectors of leishmaniasis. Emerging Infectious Diseases 7: 872-876.

47.Killick-Kendrick, R., Killick-Kendrick, M., Focheux, C., Dereure, J., Puech, M.P. and Cadiergues, M.C. 1997. Protection of dogs from the bites of phlebotomine sandflies by deltamethrin collars for the control of canine leishmaniasis. Medical and Veterinary Entomology 11: 105-111.

48.David, J.R., Stamm, L.M., Bezarra, H.S., Nonato Souza, R., Killick-Kendrick, R. and Oliveira Lima, J.W. 2001. Deltamethrin-impregnated plastic dog collars have a potent anti-feeding effect on Lutzomyia longipalpis and Lutzomyia migonei. Memorias do Instituto Oswaldo Cruz 96: 839-847.

49.Halbig, P., Hodjati, M.H., Mazoloumi-Gavgani, A.S., Mohite, H. and Davies, C.R. 2000. Further evidence that deltamethrin-impregnated collars protect dogs from sandfly bites. Medical and Veterinary Entomology 14: 223-226.

50.Maroli, M., Mizzoni,V., Siragusa, C., D'Orazi A. and Gradoni, L. The impact of deltamethrin-impregnated dog collars on the incidence of canine leishmaniasis: a pilot field study in an endemic area of southern Italy. Medical and Veterinary Entomology, 15: in press.

51.Killick-Kendrick, R. 1987. Methods for the study of phlebotomine sandflies. In: The leishmaniases in biology and medicine, vol.1, (Eds Peters, W. and Killick-Kendrick, R.), London/Orlando: Academic Press 473-497.

52.Alexander, B. 2000. Sampling methods for phlebotomine sand flies. Medical and Veterinary Entomology 14: 109-122.

53.Killick-Kendrick, R., Maroli, M. and Killick-Kendrick, M. 1991. Bibliography on the colonization of phlebotomine sandflies. Parassitologia, 33 (Suppl.), 321-333.

54.Killick-Kendrick, M. and Killick-Kendrick, R. 1991. The initial establishment of sandfly colonies. Parassitologia 33 (Suppl.1): 321-333.

55.Killick-Kendrick, R. and Killick-Kendrick, M. 1987. The laboratory colonization of Phlebotomus ariasi (Diptera: Psychodidae). Annales de Parasitologie humaine et compare 52: 354-356.

56.Killick-Kendrick, M., Killick-Kendrick, R., Añez, N., Nieves, E., Scorza, J.V. and Tang, Y. 1997. The colonization of Lutzomyia youngi and the putative role of free-living nematodes in the biology of plebotomine sandfly larvae. Parasite 4: 269-271.

57.Young, D.J., Perkins, P.V. and Endris, R.G. 1981. A larval diet for rearing phlebotomine sand flies. Journal of Medical Entomology 18: 446.

58.Mauricio, I.L., Howard, M.K., Stothard, J.R. and Miles, M.A. 1999. Genomic diversity in the L. donovani complex. Parasitology 119: 237-246.

CANINE RESERVOIRS AND LEISHMANIASIS: EPIDEMIOLOGY AND DISEASE.

Lenea Maria Campino

Unidade de Leishmanioses/Centro de Malária e Outras Doenças Tropicais. Instituto de Higiene e Medicina Tropical. Universidade Nova de Lisboa. Portugal.

INTRODUCTION

Leishmaniasis is a group of diseases caused by infection with intracellular protozoa of the family Trypanosomatidae, genus *Leishmania* Ross, 1903, which are transmitted to the host by phlebotomine sand flies. One subset, human visceral leishmaniasis (VL) is caused by parasites of the *L. donovani* complex. The dog *Canis familiaris* L., 1758 is the main reservoir for the maintenance of VL in the Palearctic and Neotropical regions. Canine leishmaniasis (CanL) caused by one species of the complex, *Leishmania* (*Leishmania*) *infantum* Nicolle, 1908, is endemic in all Mediterranean countries and it is also present in other regions worldwide (Table1).

A reservoir is an animal that serves as the source of human infection. According to Bray (1), a good reservoir should be in close contact with man via the sand fly, it should be susceptible to the pathogenic agent and it should make it available to the vector in sufficient quantities and in the correct state to cause infection. A good reservoir should be the principal meal source for the sand fly and both should rest and breed in the same habitat. Disease should present a chronic evolution allowing the animal to survive at least until the next transmission season. The dog is, therefore, an example of a good reservoir for *Leishmania*. The fatal nature of the canine disease suggests that the dog is a recent host in evolutionary terms. CanL was first described by Nicolle and Comte (1908), in Tunisia. The dog was the first animal to be found naturally infected and, in the following years, other cases were described in several countries of the Mediterranean basin. CanL is very important due to its high prevalence and wide geographical distribution. In some localities of Southern Europe and Northeast Brazil, CanL reaches a prevalence of nearly 40% (2, 3). Although this parasitic infection gives rise to important public health problems, control methods have been ineffective.

	Country	Leishmania
	Albania	L. infantum MON-1
	B-Herzegovina	Leishmania spp.
	Cyprus	L. infantum MON-1
	France	L. infantum MON-1
	Greece	L. infantum MON-1
	Italy	L. infantum MON-1, MON-72
Europe	Malta	L. infantum MON-1
	Portugal	L. infantum MON-1
	Spain	L. infantum MON-1, MON-105
		L. infantum MON-199
		L. infantum MON-77
	Turkey	L. infantum MON-1
	ex-USSR	L. infantum MON-?
	ex-Yugoslavia	L. donovani s.l.
	Algeria	L. infantum MON-1, MON-77, MON-34
	Egypt	L. infantum MON-98
Africa	Morocco	L. infantum MON-1
	Senegal	L. infantum MON-?
	Sudan	L. infantum MON-267, L. archibaldi MON-82, MON-257
	Tunisia	L. infantum MON-1
	China	L. infantum MON-?
	Israel	L. infantum MON-?
	Lebanon	L. infantum MON-?
	Pakistan	L. infantum MON-1
Asia	Saudi Arabia	L. infantum MON-1
	Syria	L. infantum MON-1
	Yemen	L. infantum MON-1
	Brazil	L. braziliensis, L. chagasi [1]
	Bolivia	L. chagasi [1]
	Colombia	L. chagasi [1], L. braziliensis
America	Peru	L. braziliensis, L. peruviana
	USA	L. chagasi [1]
	Venezuela	L. chagasi [1], L. braziliensis

Table 1. *The dog as reservoir: proven or prime suspect* [1]
L. chagasi is identical to *L. infantum* (26)

EPIDEMIOLOGY
Focal Distribution of Cases

Leishmania infantum is the etiological agent for the zoonotic form of VL that has the dog as main reservoir. *L. infantum* is also responsible for human cutaneous leishmaniasis (CL) in the Mediterranean basin. CanL distribution is not uniform in endemic regions, but, instead, clearly focal. Canine seroprevalence varies enormously between contiguous areas (4, 5). The distribution of positive dogs in endemic areas mostly follows the distribution of the vectors (3, 6) which is also not uniform. The focal nature may thus depend on the existence of specific ecological factors that meet the biological requirements of each species of sand fly. Indeed, some authors have verified that the level of exposure to sand flies could modulate the seroconversion rates in dogs (7). Prevalence rates also appear to fluctuate over time. Studies on dog cohorts have shown large variations in prevalence that may be due to a number of factors, including human intervention (e.g. elimination or treatment of infected dogs) and natural alterations in vector populations. Alterations of ecological and demographic factors are more apparent in tropical America. The massive destruction of primary forests with concomitant development of new farmlands and rural settlements, has lead to the presence of large populations of sand fly vectors, along with dogs and foxes, both of which serve as main reservoirs. In Brazil during the past two decades, there has been massive migration of people from rural to suburban areas of large cities. VL outbreaks or new endemic areas in periurban settings are associated with the urbanization or domestication of natural zoonotic foci (8). In other areas such as around large European cities like Lisbon, Madrid and Athens, the urbanization of leishmaniasis is probably due to the flux of inhabitants from the city center to the periphery with proliferation of one-family homes with gardens where dogs are commonly kept. These settlements also provide an excellent habitat for the vector. Some authors have found higher prevalences of CanL in urban or periurban settings than in rural areas (6, 8, 9). However, results of several surveys have shown that CanL is still more common in rural than in urban areas. This fact is probably related to zoophilic preferences of the vector, in a process called "trophic deviation in rural area" (10).

It is not possible to compare prevalence rates from reports that have used different types of methods to detect infection. Early epidemiological studies were based in direct parasitological tests, which were later replaced by more sensitive serological tests, and, more recently, molecular methods. In addition, infection rates obtained by means of passive detection cannot be compared with those obtained from house to house surveys.

Breeds, Age and Sex

Although it seems that the type of immune response is genetically controlled, the breeds of dogs that are the most susceptible to the overt disease is still unknown. Epidemiological surveys have found German shepherds, Dobermans and Boxers to be the most affected breeds (11-13). However, most of these animals are used as guard dogs and, due to an "outdoor lifestyle", exposure to sand flies is expected to be greater than that of dogs kept indoors. Differences in the infection rate of a particular breed in rural versus urban areas, may be ascribed to different uses with a main outdoor activity component versus companionship, reflective of an indoor environment. It has been suggested that dogs belonging to longhaired breeds show a lower prevalence than shorthaired breeds (14), although sand flies prefer to feed on the internal face of the ears and nose of dogs.

Prevalence of the disease in dogs under one year of age is very low due to the long incubation period and where transmission is seasonal. Epidemiological studies have shown that the prevalence of infection is highest in dogs up to three years old and that it can be related to the time of exposure to phlebotomine activity (15, 11). The prevalence of infection in dogs is not sex-related (9, 14, 4).

Wild Canidae

A common feature of *L. infantum* foci is that the dog is almost invariably the peridomestic reservoir of the parasite, although in some regions other mammals such as other canids and rodents have also been incriminated, principally as wild reservoirs. Several types of canids often found near human environments, have been found to be parasitized, thus serving as potential wild reservoirs for VL (reviewed in16). The first was the jackal *Canis aureus* L. 1758 found in Tadjikstan, in 1946. The wolf *Canis lupus* L., 1758 was detected parasitized in Central Asia, and in China, raccoon dogs *Nyctereutes procynoides* Temminck, 1839 are suspected of being the wild reservoir. Among wild Canidae, the fox is the most representative, with several species and genera being affected. In Brazil, some foxes (*Lycalopex vetulus* Burmeister, 1854) were found infected in Ceara as well as foxes (*Cerdocyon thous* L., 1766) in the Amazonian and southeastern regions. The existence of an autonomous or semi-autonomous sylvatic cycle in the Mediterranean basin was proposed after two foxes (*Vulpes vulpes* L., 1758) were found infected in the south of France, which was supported by later findings of other infected foxes in Italy, Spain and Portugal. A recent study performed in a Brazilian endemic area suggests that foxes do not maintain a transmission cycle independent of infectious dogs, but rather they suffer spill-over infection from these animals (17). Both stray and feral dogs may have an important role in the spread of *Leishmania.*

Stray dogs usually suffer from malnutrition and are consequently more susceptible to disease.

Accidental Hosts

As mentioned before, a reservoir is regarded as the system in which the parasite population is maintained indefinitely. Thus, one cannot incriminate an animal as a reservoir only by its susceptibility to the parasite. Animals found infected once can be accidental or occasional hosts in which infection develops under unusual conditions, often as a terminal step in the transmission cycle. However, these animals may introduce or reintroduce the infection into parasite free areas and may thus play a role in the epidemiology of leishmaniasis. The dog is also a host for species other than *L. infantum*, such as *L. arabica*, *L. major*, *L. tropica*, *L. mexicana* and *L.(Viannia)* spp. While the role of the dog as a reservoir of *L. infantum* is confirmed in many parts of the world, its role in the maintenance of the life cycle of other *Leishmania* species is not completely clear. Nevertheless, there is increasing evidence that the dog is the domestic reservoir of *L.(V.) braziliensis* and *L.(V.) peruviana* (18). In most other cases, the dog seems to be an accidental host.

Old World

Amongst the VL endemic areas of the Old World such as India, parts of China, the Sudan and Kenya, there are variations in the importance of the dog in transmission. In the Indian subcontinent, VL caused by *L. donovani* is an anthroponose and transmission is person to person via the vector. Whereas anthroponotic transmission is prevalent in eastern China, in the central and northwestern regions the dog is the reservoir. Recently, a canine seroprevalence of 43% was found in an endemic focus of VL in eastern Sudan. Some parasite cultures were indistinguishable by isoenzyme typing from those isolated from human cases occurring in that focus, suggesting that the dog acts as a reservoir (19). In Kenya, dogs are probably accidental hosts, since only, few dogs selected from VL patient's homes, were found to be infected (20).

Most recently, *L. infantum* has been incriminated as the only causative agent of CL in Southwestern Europe. In North Africa where *L. tropica* and *L. major* are responsible for CL, some isolates from cutaneous lesions have been also identified as *L. infantum* (21). *L. infantum* zymodeme MON-1 was the most common agent of CanL and VL. CL is mainly caused by dermotropic zymodemes (MON-11, MON-24, MON-29, MON-33, MON-78, MON-111). Although there is no doubt that the dog is the reservoir for zymodeme MON-1, the role of the dog in the transmission of dermotropic zymodemes is not clear.

Leishmania Zymodemes other than MON-1 (MON-34, MON-72, MON-82, MON-98, MON-105, MON-199, MON-257, MON-267) have been occasionally isolated from dogs in Spain, Italy, Algeria, Egypt and Sudan (22, 23, 24, 19). The concomitant presence of two zymodemes of *L. infantum* (MON-1 and MON-77) was found in one dog from Catalonia, Spain (25).

New World

Human leishmaniasis in the neotropics is zoonotic. The human visceral form is caused by *L. chagasi*, which is synonymous with *L. infantum* suggesting introduction of *L. chagasi* into the New World in recent history (26). The dog is considered to be the main reservoir and CanL is assuming increasing importance in many countries of the Americas. Although VL is present in various Latin American countries, more than 90% of all cases occur in Brazil, especially in the northeast. The seroprevalence of canine infection in these endemic regions is also high. Evans and others (3) found a prevalence of 38% in a rural area of the Brazilian State of Ceara, and, although the disease used to be predominantly rural, it is increasing in urban/suburban settings.

Human American cutaneous leishmaniasis (ACL) occurs between the southern part of the USA and northern Argentina, and it is thus an important public health problem in Latin America. *Leishmania (Viannia) braziliensis* causes the most important and severe form of ACL. In endemic areas, dogs have frequently been found infected with parasites of the *L. braziliensis* complex, providing further evidence of their role as reservoir of ACL (18). Originally associated with forest settlements, *L. braziliensis* has now adapted to the domestic environment due to deforestation and urbanization.

Relationship Between Human Visceral Leishmaniasis and Canine Leishmaniasis

Despite some studies showing a direct relationship between the prevalence of leishmaniasis in the canine and human populations (27), CanL is much more prevalent and more widely distributed than VL and it does not strongly correlate with prevalence in humans. In Mediterranean countries, there are foci where the prevalence of canine infection is high, but VL is hypoendemic or sporadic. Furthermore, in some regions VL is apparently unknown but CanL prevalence is very high. This happens in Senegal (28), in the Tours region of France (29) and in the island of Ustica, Italy where canine prevalence reaches 37% (2). Very rare sporadic cases of VL had been reported in the USA, however, since January 2000, hunting dogs from 21 states in the USA and in Ontario, Canada, were found to be infected with

Leishmania. From almost 11,000 fox-hunting dogs screened for specific antibodies, about 12% were seropositive (30). In Brazilian endemic areas, canine infection is more frequent than in humans (4).

Some authors have verified that the systematic removal of *Leishmania*-infected dogs from endemic areas, as carried out in Sicily, Middle Asia, China and Jacobina, Brazil has brought a marked decrease in the incidence of human cases (31-33). However, Dietze and others (34) attempted a similar endeavor in Espirito Santo, Brazil but they observed no significant decrease in the incidence of human infection.

Direct Transmission

The transmission of *Leishmania* among dogs by direct contact (without intervention of the sand fly) has been admitted by the report of autochthonous CanL in Northern European countries, where the existence of sand fly vectors is unknown. Direct transmission of the infection between co-housed dogs or from a bitch to a puppy has been described (35). In the USA and in Canada, the infection of a large number of dogs that has been recently described was suggested to have occurred by direct contact. Direct transmission has been shown to occur, however the high prevalence found in this case is not easily explained.

Asymptomatic Carriers

Infected asymptomatic dogs were not considered to be infective or to be only slightly infective and symptomatic dogs were able to infect a large proportion (70-80%) of the vector population (36). More recently, it has been suggested that asymptomatic dogs could transmit the infection to the sand fly, causing infection rates comparable to those obtained in phlebotomines feeding on symptomatic dogs (37). Furthermore, in symptomatic dogs, parasites were more frequently detected in cutaneous lesions than in intact skin. Thus, it should be considered that despite the fact that symptomatic dogs are the most infective to sand flies, asymptomatic animals are still infective to large numbers of sand flies. Although the demonstration of parasites in infected dogs increase with antibody levels and severity of clinical signs, serological data is not a reliable indicator of presence of infection. The existence of asymptomatic parasitized dogs with or without antibodies was observed in natural and experimental canine infections (37, 38). Seroepidemiological surveys performed on CanL foci of Europe have revealed that more than half of the dogs with anti-*Leishmania* antibodies are asymptomatic (11-13). It has already been demonstrated that the real rate of infection is higher than the one estimated using classical serological tests. Studies using PCR or immunoblotting performed in dogs living in endemic areas found higher prevalences than with the classical tests (39).

In conclusion, it is important to detect all infected dogs and to understand the role of asymptomatic dogs as reservoirs, as they are often undetectable upon clinical examination and a large number escape control measures, thus contributing to the spread of leishmaniasis. Information on the geographical distribution and prevalence of CanL is essential to the design and implementation of appropriate control measures.

THE DISEASE: Viscerocutaneous Leishmaniasis
Clinical Signs

Not all dogs inoculated with promastigotes by the vector develop clinical signs. Nowadays, it is recognized that asymptomatic infections are much more frequent than symptomatic ones. However, *L. infantum* infection can cause a severe systemic disease in dogs. Visceral leishmaniasis is characterized by lesions in internal organs, although in dogs and foxes, mucocutaneous pathology is also present. Lymphoid organs, skin, liver and kidneys are the most affected, characterized by a chronic and proliferative inflammatory process with profuse infiltrates of macrophages, lymphocytes and plasma cells. These inflammatory infiltrates may be diffuse or granulomatous, both of which can be associated with degeneration and necrosis. The broad spectrum of clinical and laboratory features depends on the phase of the infection. In fact, we can distinguish two clinical forms, the patent and the latent, the latter developing to overt disease or to self-cure (36). The conversion of seropositive to seronegative status of self-cured cases has been observed in natural and experimental infections (36, 15, 40).

Classically, leishmaniasis has a long, albeit variable, incubation time followed by an insidious onset. After seroconversion, dogs may remain asymptomatic for long periods, usually from one to 12 months, before clinical signs develop. In the natural infection, however, it is not possible to precisely determine the duration of the incubation period, since the time of inoculation is always unknown. Long periods without clinical signs after infection have been observed in the experimental canine model. Indeed, it has been demonstrated that infected dogs can remain asymptomatic for periods as long as 25 months after inoculation (41). These states may correspond to the incubation period before clinical active disease (i.e. pre-patent period) appears or to resistant cases and it is possible that a significant fraction of infected dogs may never reveal clinical signs or specific antibodies (42, 38).

Once the disease becomes patent it can rapidly progress to death within weeks or months, or more frequently, to a chronic course lasting several years. The initial period may be accompanied by non-specific and moderate manifestations, developing to marked and characteristic clinical signs later on. Skin abnormalities are the most usual manifestations of CanL. Alopecia around the eyes, on the ears and on the sacrolumbar region, dry exfoliative dermatitis in glabrous skin and hypertrophy and exaggerated nail growth (onychogryphosis) are common. Superficial ulcerative lesions that are crusted and bleed easily on contact, are frequent on head, ears, muzzle, periocular region and limbs. Other ulcerative lesions (ulcers) are regularly shaped, deep, have red borders, a pink base and they mainly affect the anterior face of the limbs and feet joints. Ulceration is usually related to the direct action of the parasite, although it may also be attributed to necrotizing vasculitis caused by deposition of immune complexes (43). Weight loss is almost a constant finding, although anorexia is rare. Fever is occasional, as body temperature is generally below 103.1 °F (39.5°C). Pallor of mucous membranes is common. Erosions of oral and nasal mucous membranes are also frequent. Ocular manifestations such as conjuntivitis, interstitial keratitis and chorioretinitis, can also be observed. Five to 10% of dogs have episodes of epistaxis, which is sometimes the first clinical sign of the infection. The visceral manifestations are basically a generalized lymphadenomegaly and hepatosplenomegaly. Gastrointestinal signs are not frequent. In the terminal period, all organs may be invaded by the parasite, which is the probable cause of vital organs failure. Hair loss and ulceration are severe and widespread. Muscular weakness is accentuated. In this period, the wasting process is progressive and followed by cachexia. Neurologic manifestations may occur with impairment of motor function and paralysis. Renal involvement is not rare; immune-complex glomerulonephritis and tubulointerstitial nephritis have been incriminated as the main cause of the asymptomatic proteinuria, nephrotic syndrome or chronic renal failure, seen in this parasitosis (44).

In summary, the main clinical manifestations observed are: skin lesions and onychogryphosis, loss of weight, lymphadenopathy, ocular lesions, epistaxis, muscular weakness, anemia and renal failure.

Hematological, Biochemical and Immunological Parameters

Serum alterations are always present in CanL, especially a marked increase of globulins and inversion of the albumin:globulins ratio. Electrophoresis of proteins reveals a significant decrease in albumin and a combined increase in beta- and gamma-globulins, which are both characteristic but non-specific. The hematocrit is low. The moderate anemia observed is normochromic and normocytic. Leukopenia and thrombocytopenia are rare.

Clinical diagnosis must be confirmed or assessed through laboratory tests. The methodology used in the laboratory diagnosis of CanL is similar to that used for human leishmaniasis, i.e., serological tests (immunofluorescence, counterimmunoelectrophoresis, direct agglutination and enzyme-linked immunosorbent assay) for detection of antibodies and parasite detection in affected tissues (through microscopy and cultures). More sensitive techniques such as immunoblotting and DNA tests (polymerase chain reaction, probes) are now available, but the results obtained are not yet those expected.

The increased production of immunoglobulins is non-protective, and is potentially damaging. The presence of an antibody response indicates exposure to parasites, but not necessarily active disease. Serological tests remain positive for long periods of time, many months or years after clinical cure. Thus, these methods are not adequate for assessment of cure and follow-up of treated dogs. High levels of antibodies are related with advanced disease, and higher levels
occur in symptomatic as opposed to asymptomatic dogs. There is a general correlation between high antibody levels, severity of the clinical signs and high parasite load (15).

Dogs were previously thought to have no ability to develop T cell-mediated immunity but some authors have demonstrated the existence of cellular responses in naturally and experimentally infected dogs (40, 45). It was postulated that strong cellular immune responses could be associated with resistant dogs, which had no specific antibodies while symptomatic or susceptible animals were seropositive but failed to respond to parasite antigen in cell-mediated assays (42). However,
others have not verified this association (46). T cells from asymptomatic infected dogs were able to produce significant levels of IL-2, TNF and IFN-γ in response to *Leishmania* antigen compared with those from either symptomatic or uninfected dogs (42).

Importance of the Dog as Reservoir for American Cutaneous Leishmaniasis

Canine infection with dermotropic *Leishmania* species of the New World is not as well known as that caused by *L. infantum*. Many authors described the disease as a strictly mucocutaneous pathology, although no mention was made of visceral examination. In fact, it seems that this disease tends to have striking mucocutaneous manifestations. However, Reithinger and others (18) have detected parasite DNA in internal organs and in blood of dogs infected by *L. braziliensis*, suggesting hematogenous dissemination. Furthermore, *L. (Viannia)* spp have been found in viscera of other mammals, such as sloths and marsupials (47). Thus, dissemination to viscera may be more common than expected.

ACKNOWLEDGMENTS

I want to express my deep gratitude to Prof. P. Abranches, my mentor and teacher in this field, for his contribution in this project. I also would like to thank S. Cortes and I. Maurício for all the precious help they have given in the elaboration of this text.

REFERENCES

1. Bray RS. 1982. The zoonotic potential of reservoirs of leishmaniasis in the Old World. Ecol Dis 1: 257-267.
2. Mansueto S, Di Leo R, Miceli MD and Quartararo P. 1982. Canine leishmaniasis in three foci in Western Sicily. Trans R Soc Trop Med Hyg 76: 565-566.
3. Evans TG, Teixeira MJ, McAuliffe IT, Vasconcelos IAB, Vasconcelos AW, Sousa AQ, Lima JWO and Pearson RD. 1992. Epidemiology of visceral leishmaniasis in Northeast Brazil. J Inf Dis 166: 1124-1132.
4. Paranhos-Silva M, Freitas L, Santos WC, Grimaldi Jr G, Pontes-de-Carvalho LC and Oliveira-dos-Santos AA. 1996. Cross-sectional serodiagnostic survey of canine leishmaniasis due to *Leishmania chagasi*. Am J Trop Med Hyg 55: 39-44.
5. Deplazes P, Grimm F, Papaprodromou M, Cavaliero T, Gramicia M, Christofi G, Christofi N, Economides P and Eckert J. 1998. Canine leishmaniosis in Cyprus due to *Leishmania infantum* MON-1. Acta Trop 71: 169-178.
6. Tselentis Y, Gikas A and Chaniotis B. 1994. Kala-azar in Athens basin. Lancet 343: 1635.
7. Zaffaroni E, Rubaudo L, Lanfranchi P and Mignone W. 1999. Epidemiological patterns of canine leishmaniosis in Western Liguria (Italy). Vet Parasitol 81: 11-19.
8. Cunha S, Freire M, Eulalio C, Cristosvao J, Netto E, Johnson Jr W, Reed SG and Badaró R. 1995. Visceral leishmanaisis in a new ecological niche near a major metropolitan area of Brazil. Trans R Soc Trop Med Hyg 89: 155-158.
9. Amela C, Mendez I, Torcal JM, Medina G, Pachón I, Cañavete C and Alvar J. 1995. Epidemiology of canine leishmaniasis in the Madrid region, Spain. Eur J Epidiemol 11: 1-5.
10. Abranches P, Lopes F, Silva FMC, Ribeiro MMS and Pires CA. 1983. Le Kala-azar au Portugal. III. Résultats d'une enquête sur la leishmaniose canine réalisée dans les environs de Lisbonne. Comparaison des zones urbaines et rurales. Ann Parasitol Hum Comp 58: 307-315.
11. Abranches P, Silva-Pereira MCD, Conceição-Silva FM, Santos-Gomes G and Janz JG. 1991. Canine leishmaniasis: pathological and ecological factors influencing transmission of infection. J Parasitol 77: 557-561.
12. Campino L, Capela MJR, Maurício I and Ozensoy S. 1995. Abranches P. O Kala-azar em Portugal. IX. A região do Algarve: Inquérito epidemiológico sobre o reservatório canino na região de Loulé. Rev Port Doenças Infecciosas 18: 189-194.
13. Sideris V, Papadopoulou G, Dotsika E and Karagouni E. 1999. Asymptomatic canine leishmaniasis in Greater Athens area, Greece. Eur J Epidem 15: 271-276.
14. Morillas F, Rabasco FS, Ocaña J, Martin-Sanchez J, Ocaña-Wihelmi J, Acedo C and Sanchiz-Marin MC. 1996. Leishmaniosis in the focus of Axarquia region, Malaga province, Southern Spain: a survey of the human, dog and vector. Parasitol Res 82: 569-570.
15. Pozio E, Gradoni L, Bettini S and Gramiccia M. 1981. Leishmaniasis in Tuscany (Italy). V. Further isolation of *Leishmania* from *Rattus rattus* in the province of Grosseto. Ann Trop Med Parasitol 75: 393-395.

16. Abranches P. 1989. Reservoirs of visceral leishmaniasis. In Leishmaniasis. The Current Status and New Strategies for Control. Ed. D.T. Hart, Plenum Press, New York, pp. 61-69.

17. Courtenay O, Quinnell RJ and Dye C. 2001. The role of foxes (Carnivora: canidae) in the maintenance and transmission of *Leishmania infantum*: implications for peridomestic control. Worldleish 2; Crete, Greece, Abstract Book p. 38.

18. Reithinger R, Lambson B, Barker D and Davies C. 2000. Use of PCR to detect *Leishmania (Viannia)* spp. In dog blood and bone marrow. J Clin Microbiol 38: 748-751.

19. Dereure J, Boni M, Pratlong F, El Hadi M, Osman OF, Bucheton B, El-Safi S, Feugier E, Musa MK, Davoust B, Dessein A and Dedet JP. 2000. Visceral leishmaniasis in Sudan: first identifications of *Leishmania* from dogs. Trans R Soc Trop Med Hyg 94: 154-155.

20. Mutinga MJ, Ngoka JM, Schnur LF and Chance ML. 1980. The isolation and identification of leishmanial parasites from domestic dogs in the Machakos District of Kenya, and the possible role of dogs as reservoirs of Kala-azar in East Africa. Ann Trop Med Parasit 74: 139-144.

21. Gramicia M, Ben-Ismail R, Gradoni L, Ben-Rachid MS and Ben-Said M. A. 1991. *Leishmania infantum* enzymatic variant, causative agent of cutaneous leishmaniasis in North Tunisia. Trans R Soc Trop Med Hyg 85: 370-371.

22. Gramicia M, Gradoni L, di Martino L, Romano R and Ercolini D. 1992. Two syntopic zymodemes of *Leishmania infantum* cause human and canine visceral leishmaniasis in the Naples area, Italy. Acta Trop 50: 357-359.

23. Martin-Sanchez J, Morillas-Marquez F, Sanchiz-Marin MC and Acedo-Sanchez C. 1994. Isoenzimatic characterization of the etiologic agent of canine leishmaniasis in the Granada region of southern Spain. Am J Trop Med Hyg 50: 758-762.

24. Harrat Z, Pratlong F, Belazzoug S, Dereure J, Deniau M, Rioux JA, Belkaid M and Dedet JP. 1996. *Leishmania infantum* and *L. major* in Algeria. Trans R Soc Trop Med Hyg 90: 625-629.

25. Pratlong F, Portus M, Rispail P, Moreno G, Bastien P and Rioux JA. 1989. Présence simultanée chez le chien de deux zymodemes du complexe *Leishmania infantum*. Ann Parasitol Hum Comp 64: 312-314.

26. Mauricio I, Stothard JR and Miles MA. 2000. The strange case of *Leishmania chagasi*. Parasitol Today 16: 188-189.

27. Marty P, Le Fichoux Y, Giordana D and Brugnetti A. 1992. Leishmanin reaction in the human population of a highly endemic focus of canine leishmaniasis in Alpes-Maritimes, France. Trans R Soc Trop Med Hyg 86: 249-250.

28. Ranque P,Bussieras 1971. J. La leishmaniose canine au Sénégal. Med Afrique Noire 18:761-762.

29. Houin R, Jolivet G, Combescot C, Deniau M, Puel F, Barbier D and Romano P. 1977. Kerbeuf D. Étude préliminaire d'un foyer de leishmaniose canine dans la région de Tours. Colloques Internationaux du C.N.L.S. 239: 109-115.

30. Enserink M. 2000. Has Leishmaniasis become endemic in the U.S.? Science 290: 1881-1883.

31. Lysenko AJ. 1971. Distribution of leishmaniasis in the Old World. WHO 44: 515-520.

32. Gradoni L, Gramicia M, Mancianti F and Pieri S. 1988. Studies on canine leishmaniasis control. 2. Effectiveness of control measures against canine leishmaniasis in the Isle of Elba, Italy. Trans R Soc Trop Med Hyg 82: 568-571.

33. Asford DA, David JR, Freire M, David R, Sherlock I, Eulálio MC, Sampaio DN and Badaró R. 1998. Studies on control of visceral leishmaniasis: impact of dog control on canine and human visceral leishmanaisis in Jacobina, Bahia, Brazil. Am J Trop Med Hyg 59: 53-57.

34. Dietze R, Falqueto A, Valli L, Rodrigues T, Boulos M and Corey R. 1997. Diagnosis of canine visceral leishmaniasis with a dot-enzyme-linked immunosorbent assay. Am J Trop Med Hyg 53: 40-42.

35. Mancianti F, Sozzi S. 1995. Isolation of *Leishmania* from a newborn puppy. Trans R Soc Trop Med Hyg 89: 402.

36. Lanotte G, Rioux JA, Perieres J and Vollhardt Y. 1979. Écologie des leishmanioses dans le sud de la France. 10. Les formes évolutives de la leishmaniose viscérale canine. Elaboration d'une tipologie bioclinique à finalité épidemiologique. Ann Parasitol 54: 277-295.

37. Molina R, Amela C, Nieto J, San-Andrés M, González F, Castillo JA, Lucientes J and Alvar J. 1994. Infectivity of dogs naturally infected with *Leishmania infantum* to colonized *Phlebotomus perniciosus*. Trans R Soc Trop Med Hyg 88: 491-493.

38. Campino L, Santos-Gomes G, Riça-Capela MJ, Cortes S and Abranches P. 2000. Infectivity of promastigotes and amastigotes of *Leishmania infantum* in a canine model for leishmaniosis. Vet Parasitol 92: 269-275.

39. Solano-Gallego L, Morell P, Arboix M, Alberola J and Ferrer L. 2001. Prevalence of *Leishmania infantum* infection in dogs living in an area of canine leishmaniasis endemicity using PCR on several tissues and serology. J Clin Microbiol 39: 560-563.

40. Abranches P, Santos-Gomes G, Rachamim N, Campino L, Schnur LF and Jaffe CL. 1991. An experimetal model for canine visceral leishmaniasis. Parasite Immunol 13: 537-550.

41. Oliveira GGS, Santoro F and Sadigursky M. 1993. The subclinical form of experimental visceral leishmaniasis in dogs. Mem Inst Oswaldo Cruz 88: 243-248.

42. Pinelli E, Gonzalo R, Boop C, Rutten V, Gebhard D, Real G and Ruitenberg E. 1995. *Leishmania infantum*-specific T cell lines derived from asymptomatic dogs that lyse infected macrophages in a Major Histocompatibility Complex-restricted manner. *Eur J Immunol* 25: 1594-1600.

43. Pumarola M, Brevik L, Badiola J, Vargas A, Domingo M and Ferrer L. 1999. Canine leishmaniasis associated with systemic vasculitis in two dogs. J Comp Pathol 105: 279-286.

44. Koutinas AF, Kontos V, Kaldrimidou H and Leklas S. 1994. Canine leishmaniasis-associated nephropathy: a clinical, clinicopathologic and pathologic study in 14 spontaneous cases with proteinuria. Bull HVMS 45: 131-140.

45. Cabral M, O'Grady J, Gomes S, Sousa J, Thompson H and Alexander J. 1998. The immunology of canine leishmaniosis: strong evidence for a developing disease spectrum from asymptomatic dogs. Vet Parasitol 76: 173-180.

46. Leandro C, Santos-Gomes GM, Campino L, Romão P, Cortes S, Rolão N, Gomes-Pereira S, Riça-Capela MJ and Abranches P. 2001. Cell mediated immunity and specific IgG1 and IgG2 antibody response in natural and experimental canine leishmaniosis. Vet Immunol Immunopathol 6433: 1-12.

47. Lainson R. 1983. The American leishmaniases: some observations on their ecology and epidemiology. Trans R Soc Trop Med Hyg 77: 569-596.

THE JOURNEY OF *LEISHMANIA* PARASITES WITHIN THE DIGESTIVE TRACT OF PHLEBOTOMINE SAND FLIES

Shaden Kamhawi.

Laboratory of Parasitic Diseases; NIAID, NIH, Bethesda, MD 20892

INTRODUCTION

Leishmaniasis is a vector-borne disease caused by *Leishmania* parasites and transmitted exclusively by the bite of infected phlebotomine sand flies. In humans, leishmaniasis results in a spectrum of clinical manifestations in which the parasite, sand fly vector and host immune system act in concert to determine the outcome of disease. Sand flies are the driving force in the spread of leishmaniasis and prevalence of a *Leishmania* species is entirely governed by the availability of competent vectors. This lends itself to the question "what is a competent vector?" The vectorial competence of sand flies for a particular *Leishmania* species is complex and multi-factorial. Though extrinsic factors such as geographical distribution of the flies and their feeding preferences can limit the spread of a parasite species, vector competence is mainly determined by factors intrinsic to the fly, some of a general nature and others highly specific, that challenge the successful completion of the *Leishmania* life cycle within the digestive tract of the fly. Apart from the subgenus *Vianna*, termed peripylarian *Leishmania*, whose development includes a stage in the hindgut, all other species of *Leishmania* that produce disease in mammalian hosts are suprapylarian and confine their development to the midgut and foregut of the sand fly (1). Focusing mainly on suprapylarian *Leishmania*, this chapter reviews the advances we have made in our understanding of the perilous journey *Leishmania* parasites undertake within the digestive tract of sand flies to achieve successful transmission.

Overview of the life cycle of *Leishmania* in the sand fly

Leishmania parasites are dimorphic, existing as intracellular amastigotes within the macrophage of the mammalian host, and as flagellated promastigotes in the digestive tract of the sand fly. When a sand fly feeds on infected tissue of a mammalian host, amastigotes are ingested with the bloodmeal and pass directly into the abdominal part of the midgut. The blood stimulates midgut cells to secrete a chitinous protein matrix called the peritrophic membrane. The membrane envelops the bloodmeal within 4 hours (2) and is fully formed within 24 hours after blood ingestion (3). The peritrophic membrane protects the gut epithelium from the contents of the bloodmeal, and acts as a barrier that regulates the diffusion of digestive enzymes secreted by gut epithelial cells (3). The digestion of the bloodmeal is achieved over 4-5 days after which the undigested remnants are excreted. A successful completion of the life cycle of *Leishmania* parasites in the fly requires that they survive the digestive enzymes of the host; avoid expulsion from the gut; and, at a later stage, migrate anteriorly and break free from the midgut epithelium for transmission to the mammalian host. To accomplish this, the parasites undergo several developmental changes, each adapted to overcome one or more of these barriers. Within 24 hours of ingesting an infected bloodmeal, amastigotes transform into short ovoid promastigotes called procyclics. These forms divide in the abdominal midgut forming rosettes that move sluggishly within the bloodmeal. Around 2-3 days post feeding, procyclics transform into large slender promastigotes termed nectomonads. These forms multiply rapidly and localize at the anterior part of the abdominal midgut. The peritrophic membrane begins to disintegrate 3 days after the bloodmeal due to a chitinase secreted by the sand fly. This disintegration is aided by a parasite-derived chitinase that accelerates the exit of nectomonads into the midgut lumen of the fly (4). Nectomonads attach to the lining of the midgut epithelium, which protects them from getting excreted with the undigested remnants of the bloodmeal, and continue to divide rapidly as they resume their anterior migration towards the thoracic midgut. At around 5-6 days post feeding, the bloodmeal is completely digested and the nectomonads begin their differentiation into two main forms, the haptomonads and metacyclics. Haptomonads are highly specialized forms that appear only in the area of the stomodeal valve. They adhere to the cuticular lining of the valve (5, 6) and to one another forming a plug that appears to consist of parasites embedded in a gelatinous matrix (6-10). The plug most likely blocks the food pathway and interferes with the working of the pharyngeal and cibarial pumps, making it difficult to for the fly to engorge. This difficulty in feeding results in increased probing and promotes the transmission of metacyclics (5, 11-13). A study by Schlein et al. (4) demonstrated that a chitinase produced by the parasites damages the chitin based cuticular lining of the stomodeal valve, further disrupting the mechanics of feeding and contributing to enhanced transmission. The differentiating metacyclics are freely motile and accumulate just behind the stomodeal valve, well positioned for egestion from the mouthparts when the fly attempts to take another bloodmeal. Figure 1 illustrates the chronological appearance of the main forms of *Leishmania* developing within the digestive tract of a sand fly.

Molecules central to vector-parasite interactions during the critical stages of the *Leishmania* life cycle in the sand fly, and their relevance to vector competence are discussed below.

Surviving the digestive enzymes of the sand fly

Proteolytic enzymes secreted by the sand fly to digest the bloodmeal are harmful to the parasites. In *P. papatasi* infected with *L. major* amastigotes, 50% mortality of the parasites was observed as early as 4 hours after bloodmeal ingestion (3). Factors that limit the induction or activity of these enzymes, such as bloodmeals devoid of or containing decreasing amounts of serum, or the addition of protease inhibitors, enhance the survival of *Leishmania*, even in unnatural vector species (3, 14-16). In order to increase their chance of survival in the early midgut environment, *Leishmania* parasites appear to have evolved a specific ability to modulate the activity of midgut digestive enzymes in their natural vector species. *Leishmania major*, but not other *Leishmania* species, inhibited or delayed the peak activity of proteolytic enzymes produced by its putative vector *P. papatasi* (15, 17). Secreted glycoconjugates of *Leishmania* promastigotes appear to play a part in protecting *Leishmania* parasites from the digestive enzymes of its sand fly host. Cell surface and secreted phosphoglycan containing molecules are the major class of glycoconjugates expressed by *Leishmania* parasites. *Leishmania* phosphoglycans share a common phosphorylated disaccharide structure [-PO$_4$-6Galβ1-4Man1α-] and include the glycosylphosphatidylinositol (GPI) membrane bound lipophosphoglycan (LPG), secreted glycoproteins: acid phosphatase (AP) and proteophosphoglyans (PPG); and secreted extracellular phosphoglycans (PG) (18). Other GPI anchored glycoconjugates include the glycosylinositol phospholipids (GIPLs) and the surface protease GP63. The basic structure of the major glycoconjugates of *Leishmania* is illustrated in Figure 2.

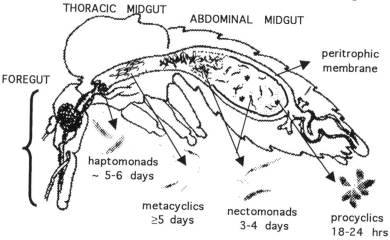

THORACIC MIDGUT

ABDOMINAL MIDGUT

peritrophic membrane

FOREGUT

haptomonads
~ 5-6 days

metacyclics
≥5 days

nectomonads
3-4 days

procyclics
18-24 hrs

Figure1. *Chronological appearance and location of the major morphological forms of Leishmania in the sand fly midgut.*

The addition of glycoconjugates from a compatible strain of *L. major*, but not *L. donovani*, enhanced the survival of a non-compatible strain of *L. major* in *P. papatasi*, and delayed bloodmeal digestion (19). The importance of secreted phosphoglycans in protection from digestive enzymes arose from the analysis of parasite mutants. Parasite mutants that are deficient in the synthesis of all phosphoglycans (*LPG2* mutants) were highly sensitive to the conditions in the early blood fed midgut, while growth of mutants whose deficiency is restricted to the synthesis of LPG (*LPG1* mutants) was only slightly reduced (20). Phosphoglycan containing molecules of different *Leishmania* exhibit interspecies differences that may affect their ability to inhibit or delay the activity of digestive enzymes. This may account for some of the specificity observed in the ability of parasites to inhibit digestive enzymes of their putative sand fly vectors. Selective inhibition of digestive enzymes, mainly shown for the *L. major-P. papatasi* pair, needs to be confirmed on a broader scale, by the demonstration of similar outcomes in other compatible parasite-vector pairs. LPG is the largest and most abundant molecule on the surface of *Leishmania* promastigotes and forms a dense gycocalyx on their surface. By virtue of this organization, and the nature of the molecule, it probably contributes, albeit to a lesser degree than secreted phosphoglycans, to protection from

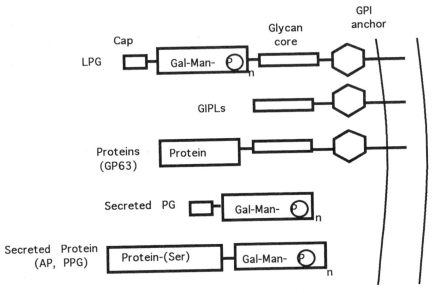

Figure 2. *The major glycogonjugates of Leishmania parasites, showing the common structures shared between membrane-bound and secreted molecules.*

digestive enzymes of the fly. Conversely, the GPI-anchored GP63 protease, also abundant on the surface of promastigotes, does not seem to play an important role in the survival of *Leishmania* promastigotes within the sand fly midgut. The deletion of GP63 genes in *L. major* did not affect its growth and development in the sand fly (21). This is not entirely surprising since LPG, by virtue of its size and abundance, was shown to effectively mask other surface anchored molecules on the promastigote (22). The peritrophic membrane also offers a degree of protection against digestive enzymes, acting as a physical barrier that slows down their diffusion into the vicinity of the parasites (3). Pimenta et al. (3) showed that the early transitional forms in the bloodmeal, as amastigotes transform into promastigotes, are highly sensitive to the hydrolytic activity of digestive enzymes and the presence of the peritrophic membrane provides the necessary time for them to complete their transformation to the relatively resistant promastigotes.

Surviving expulsion of the undigested bloodmeal

Sand flies excrete the undigested remnants of the bloodmeal, together with a disintegrating peritrophic membrane, around 4-5 days after feeding. *Leishmania* parasites developing in the midgut must avoid expulsion with the bloodmeal. To achieve this, the parasites need to escape the peritrophic membrane and anchor themselves to the midgut epithelium. In studies on the development of *L. panamensis* in *P. papatasi*, an unnatural parasite-vector pair, loss of infection was attributed to the failure of the breakdown of the peritrophic membrane (23). Further evidence that the peritrophic membrane can act as a barrier to parasite development is provided by a study where infections of *P. papatasi* with *L. major*, a natural parasite-vector pair, were lost in flies that were treated with allosamidin, an inhibitor of chitinase (3). Mainly, loss of infections of *Leishmania* species in incompetent vectors occurs at a later time point, and some studies have directly correlated the loss of infection with the loss of the bloodmeal (24-29). In nature, *P. papatasi* is only known to transmit *L. major*, even in areas where other parasite species are present. This parasite-vector specificity is mirrored in experimental studies. Infections of *P. papatasi* with *L. donovani* or *L. tropica* are lost by day 4-7 after feeding, while those with *L. major* persist in the large majority of the flies (28). It is worthwhile mentioning that the viability and growth of *L. donovani* and *L. tropica* in this study was comparable to that of *L. major* during the first few days following the ingestion of the bloodmeal, indicating that at least for these vector-parasite combinations their sensitivity/resistance to proteolysis plays a minimal role in the outcome of infection. A similar pattern of vector-parasite specificity is observed for *P. sergenti* with *L. tropica* (27, 29), but not for *P. argentipes* or *Lu. longipalpis* that appear to be permissive to the full development of many *Leishmania* species (5, 28-30).

LPG was implicated as the key molecule in the selective ability of certain *Leishmania* parasites to persist in the midgut of particular sand fly species. LPG forms a thick glycocalyx on the parasite surface with as many as 5 million copies expressed per cell (31). LPG of all *Leishmania* species share a basic structure consisting of a GPI anchor, a glycan core, and a repeating unit of phosphorylated disaccharides terminating with neutral capping sugars (Fig 3a). Many studies provide strong evidence that LPG mediates the attachment of parasites to midgut epithelial cells: In *in vitro* binding assays, purified LPG of *L. major*, *L. tropica*, and *L. donovani* bound to whole midguts of their natural vectors *P. papatasi*, *P. sergenti*, and *P. argentipes* respectively (28, 29, 32, 33); *L. major* and *L. donovani* LPG inhibited in a dose dependent manner the binding of promastigotes to midguts of *P. papatasi* and *P. argentipes* respectively (32, 33); promastigote mutants specifically deficient in LPG biosynthesis lost their ability to bind to the sand fly midgut (20, 34), and mutants with restored LPG expression regained their binding ability (20). The fact that ultrastructural studies consistently showed *Leishmania* parasites bound to the microvilli of midgut epithelial cells via their flagellum (9, 35-37), and the finding that a flagellar-specific protein from *L. major* inhibited *in vitro* binding of flagellar preparations to frozen sections of *P. papatasi* midguts (38), suggests the presence of additional receptors that probably assist in the attachment of parasites. LPG not only mediates attachment to the fly midgut but also controls species-specific vector competence by its structural polymorphisms. The LPG structure of several *Leishmania* species reveals a sophisticated level of inter- and sometimes intra-specific polymorphisms. In *L. major* and *L. tropica*, the repeating disaccharide units are extensively substituted with variable saccharide side chains terminating mostly with galactose for *L. major* (Fig. 3b) and arabinose for *L. tropica* (39, 40). In *L. mexicana* and the Indian strain of *L. donovani* the repeats are partially substituted (41, 42). In the Sudanese strain of *L. donovani* (Fig. 3d) they remain without substitution (43, 44). That the polymorphic nature of LPG largely accounts for parasite specific vector competence was demonstrated in *L. major-P. papatasi* and *L. tropica-P. sergenti* interactions. Parasites that possess unsubstituted LPGs, LPGs whose side chains do not terminate with galactose sugars, or genetically modified *L. major* mutants lacking side chains or terminal galactose sugars loose their ability to bind to the midgut of *P. papatasi* (20, 28, 34). Similarly, LPG of *L. tropica*, whose extensively branched LPG side chains terminate mainly with arabinose sugars, bound significantly to midguts of *P. sergenti* but not to midguts of *P. papatasi* and *P. argentipes* (29). Inversely, *P. argentipes* and *Lu. longipalpis*, infected in nature with Indian *L. donovani* and *L. chagasi* respectively, parasites whose LPGs show little or no side chain substitutions, are permissive to all *Leishmania* species studied (5, 28-30). The above studies gave rise to a theory that the structural polymorphisms of *Leishmania* LPGs are driven by the diversity of molecules functioning as parasite attachment sites. As such, the level of complexity of the midgut receptor would restrict their vector competence to those parasites evolving complimentary LPG structures.

There is little information on the midgut receptors of LPG. The only published report is of a microvillar protein identified from the midgut of *P. papatasi* that bound to the LPG of *L. major* (45). Lectins, proteins with sugar binding properties, which agglutinate *Leishmania* parasites, are also present in sand fly midguts (46-48), and demonstrate an affinity for LPG (49). However, these lectins appear to be secreted in response to the ingestion of a bloodmeal (50), which undermines their significance as LPG receptors. Instead, Volf et al. (48) put forward the hypothesis that sand fly lectins adversely affect the

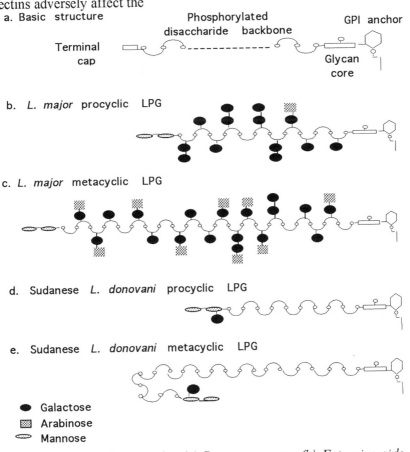

Figure 3. *The LPG molecule: (a) Basic structure; (b) Extensive side chain substitutions of L. major procyclic LPG; (c) Substitution of galactose by arabinose in L. major metacyclic LPG; (d) Unsubstituted LPG of procyclic L. donovani; (e) Elongation and folding of L. donovani metacyclic LPG.*

development of *Leishmania* in the sand fly, either by directly killing the parasites or by inhibiting their attachment to the microvillar receptor. Strong candidates as LPG receptors are a galectin gene homologue, encoding a galactose binding protein, and a C-type lectin identified exclusively from cDNA libraries of *P. papatasi* and *P. sergenti* midguts respectively (Jesus Valenzuela and Shaden Kamhawi, unpublished data). Work is presently underway to confirm the functional significance of these lectins.

Successful transmission to a mammalian host

The ultimate objective of *Leishmania* infections in the sand fly is the development of infective metacyclics, the hallmark of transmissible infections. Some of the important characteristics of *Leishmania* metacyclics are free motility, loss of agglutination with specific lectins and resistance to complement mediated-lysis and macrophage killing, all functional adaptations for transmission to and survival in the mammalian host (51-53). The differentiation of nectomonads to metacyclics is largely dependant on structural modifications of their LPG. For most *Leishmania* species studied, this modification involves a significant elongation of the LPG molecule by a 2- to 3-fold increase in the number of phosphorylated disaccharide repeats forming its backbone; and/or changes in the nature of the terminal sugars on its substituted side chains (42, 54, 55). The developmental modifications of *L. major* and *L. donovani* metacyclic LPGs are depicted in Figure 3c,e. These modifications have functional implications. The elongation of the molecules results in a significant thickening of the LPG coat covering the parasite surface. In *L major*, the thickness of the glycocalyx was shown to increase from 7nm-thick in procyclics to 17nm in their metacyclic counterparts (56). For most *Leishmania* species the elongation of metacyclic LPG is associated with complement resistance. However, as was elucidated for Sudanese *L. donovani*, the elongation of the LPG molecule is also important for altering the spatial configuration of the molecule so that the terminal binding sugars are no longer exposed and unable to mediate binding of the parasite to putative midgut receptors (33). For *L. major*, the release of attached parasites is associated with the down-regulation of side chains terminating with galactose sugars and their replacement by side chains terminating with arabinopyranose sugars (32, 54, 55). Despite the identification of several metacyclic-specific genes (57, 58), cues that induce stage differentiation of procyclics into metacyclic forms remain largely unknown. Recently, *in vitro* studies showed that tetrahydrobiopterin, a by-product of pteridine metabolism, inversely regulates differentiation into metacyclics and may be a signaling molecule for inhibition of metacyclogenesis (59). Studies are presently underway to determine whether this regulatory role is maintained in sand fly infections. Haemoglobin may be another negative regulator of metacyclogenesis. In cultures of *L. major* the addition of haemoglobin resulted in an inhibition of chitinase secretion and the formation of infective metacyclic promastigotes (60). Detachment of metacyclics frees them for transmission to the mammalian host. However, this is not sufficient for successful transmission and the synchronized formation of a biological plug appears to be essential. This plug is formed by parasites and secreted phosphoglycans that interfere with the act of feeding, increasing the need to probe and the chance for metacyclic deposition into the site of bite (5-13). One study has also demonstrated a degeneration of the stomodeal valve in *P. papatasi* infected with *L. major* and suggested that it contributes to the impairment of feeding and regurgitation of parasites (61).

Interaction between the parasites and sand fly molecules extends beyond the sand fly milieu to early events in the dermis of the mammalian host. Sand flies inoculate saliva into the wound along with metacyclics. Sand fly saliva includes an array of pharmacologically active molecules that induce vasodilatation and prevent the clotting of blood (62-64). Salivary molecules also possess immunomodulatory properties that alter host response to infection, and potentially the outcome of disease (64). Sand fly saliva was shown to exacerbate *Leishmania* infections producing significantly larger lesions and/or parasite loads compared to controls (65-70). This may be due to the inhibitory effect of sand fly saliva on macrophage functions, including antigen presentation, NO production and the ability to induce parasite primed T-cell proliferation (71, 72). Sand fly salivary molecules identified up to date as immunomodulatory include maxadilan, adenosine, hyaluronidase, adenosine deaminase and its by-product inosine, all identified from *Lu. longipalpis* and/or *P. papatasi* (64). Maxadilan, the vasodilator of *Lu. longipalpis,* is reported to inhibit T-cell activation, DTH responses, and the pro-inflammatory cytokine TNF-α, while inducing IL-6, IL-10 and prostaglandins E$_2$ (73-76). Adenosine, the vasodilator of *P. papatasi*, has anti-inflammatory properties and inhibits IL-12, IFN-γ, TNF-α and NO while enhancing the production of IL-10 (77-80). Hyaluronidase down-regulates IFN-γ, and promotes chemokine and iNOS gene expression (81, 82). Adenosine deaminase prevents T-cell apoptosis while inosine inhibits IL-12 and IFN-γ (83, 84). Apart from maxadilan, shown to exacerbate *Leishmania* infections in mice (85), the specific effects of the above mentioned immunomodulatory salivary molecules on the development of *Leishmania* parasites in their mammalian host remain to be elucidated.

Another significant consequence to the presence of saliva and metacyclics in the vicinity of the wound, is the protection observed against cutaneous leishmaniasis caused by *L. major*, as a result of pre-exposure of the host to the salivary sonicate or bites of uninfected *P. papatasi* flies (69, 86). This protection was associated with the development of a strong DTH response against salivary antigens and the production of IFN-γ at the site of bite (86). This observation has opened the door to a new approach for the development of a vaccine against *L. major* (87) and potentially other vector-borne parasitic diseases.

RECOMMENDED READING

Sacks D. and Kamhawi S. Molecular aspects of parasite-vector and vector-host interactions in leishmaniases. (Review). Annu. Rev. Microbiol. 2001, 55:453-83.

ACKNOWLEDGEMENT

I am thankful to Drs. David Sacks, Jose Ribeiro and Jesus Valenzuela for reviewing the chapter.

REFERENCES

1.Lainson, R. and J.J. Shaw. 1987. Evolution, classification and geographical distribution. The Leishmaniasis in Biology and Medicine, ed. W. Peters and R. Killick-Kendrick. Vol. 1. London: Academic Press 1-120.

2.Blackburn, K., K.R. Wallbanks, D.H. Molyneux, D.R. Lavin, and S.L. Winstanley. 1988. The peritrophic membrane of the female sandfly *Phlebotomus papatasi*. Ann Trop Med Parasitol 82: p. 613-9.

3.Pimenta, P.F., G.B. Modi, S.T. Pereira, M. Shahabuddin, and D.L. Sacks. 1997. A novel role for the peritrophic matrix in protecting *Leishmania* from the hydrolytic activities of the sand fly midgut. Parasitology 115: 359-69.

4.Schlein, Y., R.L. Jacobson, and J. Shlomai. 1991. Chitinase secreted by *Leishmania* functions in the sandfly vector. Proc R Soc Lond B Biol Sci 245: 121-6.

5.Killick-Kendrick, R., A.J. Leaney, P.D. Ready, and D.H. Molyneux. 1997. *Leishmania* in phlebotomid sandflies. IV. The transmission of *Leishmania mexicana amazonensis* to hamsters by the bite of experimentally infected *Lutzomyia longipalpis*. Proc R Soc Lond B Biol Sci 196: 105-15.

6.Walters, L.L., G.B. Modi, R.B. Tesh, and T. Burrage. 1987. Host-parasite relationship of *Leishmania mexicana mexicana* and *Lutzomyia abonnenci* (Diptera: Psychodidae). Am J Trop Med Hyg 36: 294-314.

7.Killick-Kendrick, R., K.R. Wallbanks, D.H. Molyneux, and D.R. Lavin. 1988.The ultrastructure of *Leishmania major* in the foregut and proboscis of *Phlebotomus papatasi*. Parasitol Res 74: 586-90.

8.Walters, L.L., G.L. Chaplin, G.B. Modi, and R.B. Tesh. 1989. Ultrastructural biology of *Leishmania (Viannia) panamensis* (=*Leishmania braziliensis panamensis*) in *Lutzomyia gomezi* (Diptera: Psychodidae): a natural host-parasite association. Am J Trop Med Hyg 40: 19-39.

9.Lang, T., A. Warburg, D.L. Sacks, S.L. Croft, R.P. Lane, and J.M. Blackwell. 1991.Transmission and scanning EM-immunogold labeling of *Leishmania major* lipophosphoglycan in the sandfly *Phlebotomus papatasi*. Eur J Cell Biol 55: 362-72.

10.Stierhof, Y.D., P.A. Bates, R.L. Jacobson, M.E. Rogers, Y. Schlein, E. Handman, and T. Ilg, 1999. Filamentous proteophosphoglycan secreted by *Leishmania promastigotes* forms gel-like three-dimensional networks that obstruct the digestive tract of infected sandfly gel-like three-dimensional networks that obstruct the digestive tract of infected sandfly vectors. Eur J Cell Biol 78: 675-89.

11. Beach, R., G. Kiilu, and J. Leeuwenburg. 1985. Modification of sand fly biting behavior by *Leishmania* leads to increased parasite transmission. Am J Trop Med Hyg 34: 278-82.

12. Warburg, A. and Y. Schlein. 1986. The effect of post-bloodmeal nutrition of *Phlebotomus papatasi* on the transmission of *Leishmania major*. Am J Trop Med Hyg 35: 926-30.

13. Lawyer, P.G., J.I. Githure, C.O. Anjili, J.O. Olobo, D.K. Koech, and G.D. Reid. 1990. Experimental transmission of *Leishmania major* to vervet monkeys (Cercopithecus aethiops) by bites of *Phlebotomus duboscqi* (Diptera: Psychodidae). Trans R Soc Trop Med Hyg 84: 229-32.

14. Adler, S. 1938. Factors determining the behaviour of *Leishmania* sp. in sandflies. Harefuah 14: 1-6.

15. Borovsky, D. and Y. Schlein. 1987. Trypsin and chymotrypsin-like enzymes of the sandfly *Phlebotomus papatasi* infected with *Leishmania* and their possible role in vector competence. Med Vet Entomol 1: 235-42.

16. Schlein, Y. and R.L. Jacobson. 1986. Resistance of *Phlebotomus papatasi* to infection with *Leishmania donovani* is modulated by components of the infective bloodmeal. Parasitology, Schlein, Y. and H. Romano, *Leishmania major* and *L. donovani*: effects on proteolytic enzymes of *Phlebotomus papatasi* (Diptera, Psychodidae). Exp Parasitol 62: 376-80.

18. Descoteaux, A. and S.J. 1999. Turco, Glycoconjugates in *Leishmania* infectivity. Biochim Biophys Acta 1455: 341-52.

19. Schlein, Y., L.F. Schnur, and R.L. Jacobson. 1990. Released glycoconjugate of indigenous *Leishmania major* enhances survival of a foreign L. major in *Phlebotomus papatasi*. Trans R Soc Trop Med Hyg 84: 353-5.

20. Sacks, D.L., G. Modi, E. Rowton, G. Spath, L. Epstein, S.J. Turco, and S.M. Beverley. 2000. The role of phosphoglycans in *leishmania*-sand fly interactions [In Process Citation]. Proc Natl Acad Sci U S A 97: 406-11.

21. Joshi, P.B., D.L. Sacks, G. Modi, and W.R. McMaster. 1998. Targeted gene deletion of *Leishmania major* genes encoding developmental stage-specific leishmanolysin (GP63). Mol Microbiol 27: 519-30.

22. Karp, C.L., S.J. Turco, and D.L. Sacks. 1991. Lipophosphoglycan masks recognition of the *Leishmania donovani promastigote* surface by human kala-azar serum. J Immunol 147: 680-4.

23. Walters, L.L., K.P. Irons, G.B. Modi, and R.B. Tesh. 1992. Refractory barriers in the sand fly *Phlebotomus papatasi* (Diptera: Psychodidae) to infection with *Leishmania panamensis*. Am J Trop Med Hyg 46: 211-28.

24. Heyneman, D. 1963. Leishmaniasis in the Sudan Republic. 12. Comparison of experimental *Leishmania donovani* infections in *Phlebotomus papatasi* (Diptera: Psychodidae) with natural infections found in man baited *P. orientalis* captured in a kala-azar endemic region of Sudan. Am J Trop Med Hyg 12: 725-740.

25. Lawyer, P.G., P.M. Ngumbi, C.O. Anjili, S.O. Odongo, Y.B. Mebrahtu, J.I. Githure, D.K. duboscqi and Sergentomyia schwetzi (Diptera: Psychodidae). 1990. Am J Trop Med Hyg 43: 31-43.

26. Killick-Kendrick, R. 1985. Some epidemiological consequences of the evolutionary fit between Leishmaniae and their phlebotomine vectors. Bull. Soc. Pathol. Exot. Filiales 78: 747-755.

27. Killick-Kendrick, R., M. Killick-Kendrick, and Y. Tang. 1994. Anthroponotic cutaneous leishmaniasis in Kabul, Afghanistan: the low susceptibility of *Phlebotomus papatasi* to *Leishmania tropica*. Trans R Soc Trop Med Hyg 88: 252-3.

28. Pimenta, P.F., E.M. Saraiva, E. Rowton, G.B. Modi, L.A. Garraway, S.M. Beverley, S.J. Turco, and D.L. Sacks. 1994. Evidence that the vectorial competence of phlebotomine sand flies for different species of *Leishmania* is controlled by structural polymorphisms in the

surface lipophosphoglycan. Proc Natl Acad Sci U S A 91: 9155-6.

29.Kamhawi, S., G.B. Modi, P.F. Pimenta, E. Rowton, and D.L. Sacks. 2000.The vectorial competence of *phlebotomus sergenti* is specific for *leishmania tropica* and is controlled by species-specific, lipophosphoglycan-mediated midgut attachment [In Process Citation]. Parasitology 121: 25-33.

30.Walters, L.L., K.P. Irons, G. Chaplin, and R.B. Tesh. 1993. Life cycle of *Leishmania major* (Kinetoplastida: Trypanosomatidae) in the neotropical sand fly *Lutzomyia longipalpis* (Diptera: Psychodidae). J Med Entomol 30: 699-718.

31.Sacks, D.L. 1992.The structure and function of the surface lipophosphoglycan on different developmental stages of *Leishmania promastigotes*. Infect Agents Dis 1: 200-6.

32.Pimenta, P.F., S.J. Turco, M.J. McConville, P.G. Lawyer, P.V. Perkins, and D.L. Sacks. 1992. Stage-specific adhesion of *Leishmania promastigotes* to the sandfly midgut. Science 256: 1812-5.

33.Sacks, D.L., P.F. Pimenta, M.J. McConville, P. Schneider, and S.J. Turco. 1995. Stage specific binding of *Leishmania donovani* to the sand fly vector midgut is regulated by conformational changes in the abundant surface lipophosphoglycan. J Exp Med 181: 685-97.

34.Butcher, B.A., S.J. Turco, B.A. Hilty, P.F. Pimenta, M. Panunzio, and D.L. Sacks. 1996. Deficiency in beta1,3-galactosyltransferase of a *Leishmania major* lipophosphoglycan mutant adversely influences the *Leishmania*-sand fly interaction. J Biol Chem 271: 20573-9.

35.Warburg, A., G.S. Hamada, Y. Schlein, and D. Shire. 1986. Scanning electron microscopy of *Leishmania major* in *Phlebotomus papatasi*. Z Parasitenkd 72: 423-31.

36.Molyneux, D.H. and R. Killick-Kendrick. 1987. Morphology, ultrastructure and life cycles. In The Leishmnaises in Biology and Medicine, ed. W. Peters and R. Killick-Kendrick. Vol. 1. London: Academic Press. 121-76.

37.Walters, L.L., G.B. Modi, G.L. Chaplin, and R.B. Tesh. 1989. Ultrastructural development of *Leishmania chagasi* in its vector, *Lutzomyia longipalpis* (Diptera: Psychodidae). Am J Trop Med Hyg 41: 295-317.

38.Warburg, A., R.B. Tesh, and D. McMahon-Pratt. 1989. Studies on the attachment of *Leishmania flagella* to sand fly midgut epithelium. J Protozool 36: 613-7.

39.McConville, M.J., J.E. Thomas-Oates, M.A. Ferguson, and S.W. Homans. 1990. Structure of the lipophosphoglycan from *Leishmania major*. J Biol Chem 265: 19611-23.

40.McConville, M.J., L.F. Schnur, C. Jaffe, and P. Schneider. 1995. Structure of *Leishmania lipophosphoglycan*: inter- and intra-specific polymorphism in Old World species. Biochem J 310: 807-18.

41.Ilg, T., R. Etges, P. Overath, M.J. McConville, J. Thomas-Oates, J. Thomas, S.W. Homans, and M.A. Ferguson. 1992. Structure of *Leishmania mexicana lipophosphoglycan*. J Biol Chem 267: 6834-40.

42.Mahoney, A.B., D.L. Sacks, E. Saraiva, G. Modi, and S.J. Turco. 1999. Intra-species and stage specific polymorphisms in lipophosphoglycan structure control *leishmania donovani* sand fly interactions [In Process Citation]. Biochemistry 38: 9813-23.

43.Turco, S.J., S.R. Hull, P.A. Orlandi, Jr., S.D. Shepherd, S.W. Homans, R.A. Dwek, and T.W. Rademacher. 1987. Structure of the major carbohydrate fragment of the *Leishmania donovani lipophosphoglycan*. Biochemistry 26: 6233-8.

44.Thomas, J.R., M.J. McConville, J.E. Thomas-Oates, S.W. Homans, M.A. Ferguson, P.A. Gorin, K.D. Greis, and S.J. Turco. 1992. Refined structure of the lipophosphoglycan of *Leishmania donovani*. J Biol Chem 267: 6829-33.

45.Dillon, R.J. and R.P. Lane. 1999. Detection of *Leishmania lipophosphoglycan* binding proteins in the gut of the sandfly vector. Parasitology 118: 27-32.

46.Wallbanks, K.R., G.A. Ingram, and D.H. Molyneux. 1986. The agglutination of erythrocytes and *Leishmania* parasites by sandfly gut extracts: evidence for lectin activity. Trop Med Parasitol 37: 409-13.

47.Svobodova, M., P. Volf, and R. Killick-Kendrick. 1996. Agglutination of *Leishmania promastigotes* by midgut lectins from various species of phlebotomine sandflies. Ann Trop Med Parasitol 90: 329-36.

48.Volf, P., A. Kiewegova, and M. Svobodova. 1998. Sandfly midgut lectin: effect of galactosamine on *Leishmania major* infections. Med Vet Entomol 12: 151-4.

49.Palanova, L. and P. Volf. 1997. Carbohydrate-binding specificities and physico-chemical properties of lectins in various tissue of phlebotominae sandflies. Folia Parasitol 44: 71-6.

50.Volf, P. and R. Killick-Kendrick. 1996. Post-engorgement dynamics of haemagglutination activity in the midgut of phlebotomine sandflies [published erratum appears in Med Vet Entomol 1997 Jan;11(1):104]. Med Vet Entomol 10: 247-50.

51.Sacks, D.L. and P.V. Perkins. 1984. Identification of an infective stage of *Leishmania promastigotes*. Science 223: 1417-9.

52.Howard, M.K., G. Sayers, and M.A. Miles. 1987. *Leishmania donovani* metacyclic promastigotes: transformation in vitro, lectin agglutination, complement resistance, and infectivity. Exp Parasitol 64: 147-56.

53.Puentes, S.M., D.L. Sacks, R.P. da Silva, and K.A. Joiner. 1988. Complement binding by two developmental stages of *Leishmania major* promastigotes varying in expression of a surface lipophosphoglycan. J Exp Med 167: 887-902.

54.Sacks, D.L., T.N. Brodin, and S.J. Turco. 1990. Developmental modification of the lipophosphoglycan from *Leishmania major* promastigotes during metacyclogenesis. Mol Biochem Parasitol 42: 225-33.

55.McConville, M.J., S.J. Turco, M.A. Ferguson, and D.L. Sacks. 1992. Developmental modification of lipophosphoglycan during the differentiation of *Leishmania major* promastigotes to an infectious stage. Embo J 11: 3593-600.

56.Pimenta, P.F., R.P. da Silva, D.L. Sacks, and P.P. da Silva. 1989. Cell surface nanoanatomy of *Leishmania major* as revealed by fracture- flip. A surface meshwork of 44 nm fusiform filaments identifies infective developmental stage promastigotes. Eur J Cell Biol 48: 180-90.

57.Coulson, R.M. and D.F. Smith. 1990. Isolation of genes showing increased or unique expression in the infective promastigotes of *Leishmania major*. Mol Biochem Parasitol 40: 63-75.

58.McKean, P.G., P.W. Denny, E. Knuepfer, J.K. Keen, and D.F. Smith. 2001. Phenotypic changes associated with deletion and overexpression of a stage-regulated gene family in *Leishmania*. Cell Microbiol 3: 511-23.

59.Cunningham, M.L., R.G. Titus, S.J. Turco, and S.M. Beverley. 2001. Regulation of tetrahydrobiopterin. Science 292: 285-7.

60.Schlein, Y. and R.L. Jacobson. 1994. Haemoglobin inhibits the development of infective promastigotes and chitinase secretion in *Leishmania major* cultures. Parasitology 109: 23-8.

61.Schlein, Y., R.L. Jacobson, and G. Messer. 1992. *Leishmania* infections damage the feeding mechanism of the sandfly vector and implement parasite transmission by bite. Proc Natl Acad Sci U S A 89: 9944-8.

62.Ribeiro, J.M. 1987. Role of saliva in blood-feeding by arthropods. Annu Rev Entomol 32: 463-78.

63.Ribeiro, J.M. 1995. Blood-feeding arthropods: live syringes or invertebrate pharmacologists? Infect Agents Dis 4: 143-52.

64.Kamhawi, S. 2000. The biological and immunomodulatory properties of sand fly saliva and its role in the establishment of *Leishmania* infections. Microbes and Infection 2: p. 1-9.

65.Titus, R.G. and J.M. Ribeiro. 1988. Salivary gland lysates from the sand fly *Lutzomyia longipalpis* enhance *Leishmania* infectivity. Science 239: 1306-8.

66.Samuelson, J., E. Lerner, R. Tesh, and R. Titus. 1991. A mouse model of *Leishmaniabraziliensis* braziliensis infection produced by coinjection with sand fly saliva. J

Exp Med 173: 49-54.
67.Theodos, C.M., J.M. Ribeiro, and R.G. Titus. 1991. Analysis of enhancing effect of sand fly saliva on *Leishmania* infection in mice. Infect Immun 59(5): p. 1592-8.
68.Lima, H.C. and R.G. Titus. 1996. Effects of sand fly vector saliva on development of cutaneous lesions and the immune response to *Leishmania braziliensis* in BALB/c mice. Infect Immun 64(12): p. 5442-5.
69.Belkaid, Y., S. Kamhawi, G. Modi, J. Valenzuela, N. Noben-Trauth, E. Rowton, J. Ribeiro, and D.L. Sacks. 1998. Development of a natural model of cutaneous leishmaniasis: powerful effects of vector saliva and saliva preexposure on the long-term outcome of *Leishmania major* infection in the mouse ear dermis. J Exp Med 188(10): p. 1941-53.
70.Mbow, M.L., J.A. Bleyenberg, L.R. Hall, and R.G. Titus. 1998. Phlebotomus papatasi sand fly salivary gland lysate down-regulates a Th1, but up-regulates a Th2, response in mice infected with *Leishmania major*. J Immunol 161(10): p. 5571-7.
71.Theodos, C.M. and R.G. Titus. 1993. Salivary gland material from the sand fly Lutzomyia longipalpis has an inhibitory effect on macrophage function in vitro. Parasite Immunol 15(8): p. 481-7.
72.Hall, L.R. and R.G. Titus.1995. Sand fly vector saliva selectively modulates macrophage functions that inhibit killing of *Leishmania major* and nitric oxide production. J Immunol 155(7): p. 3501-6.
73.Qureshi, A.A., A. Asahina, M. Ohnuma, M. Tajima, R.D. Granstein, and E.A. Lerner 1996. Immunomodulatory properties of maxadilan, the vasodilator peptide from sand fly salivary gland extracts. Am J Trop Med Hyg 54(6): p. 665-71.
74.Bozza, M., M.B. Soares, P.T. Bozza, A.R. Satoskar, T.G. Diacovo, F. Brombacher, R.G. Titus, C.B. Shoemaker, and J.R. David. 1998. The PACAP-type I receptor agonist maxadilan from sand fly saliva protects mice against lethal endotoxemia by a mechanism partially dependent on IL-10. Eur J Immunol 28(10): p. 3120-7.
75.Soares, M.B., R.G. Titus, C.B. Shoemaker, J.R. David, and M. Bozza. 1998. The vasoactive peptide maxadilan from sand fly saliva inhibits TNF-alpha and induces IL-6 by mouse macrophages through interaction with the pituitary adenylate cyclase-activating polypeptide (PACAP) receptor. J Immunol 160(4): p. 1811-6.
76.Lanzaro, G.C., A.H. Lopes, J.M. Ribeiro, C.B. Shoemaker, A. Warburg, M. Soares, and R.G. Titus. 1999. Variation in the salivary peptide, maxadilan, from species in the *Lutzomyia longipalpis* complex [In Process Citation]. Insect Mol Biol 8(2): p. 267-75.
77.Hasko, G., C. Szabo, Z.H. Nemeth, V. Kvetan, S.M. Pastores, and E.S. Vizi. 1996. Adenosine receptor agonists differentially regulate IL-10, TNF-alpha, and nitric oxide production in RAW 264.7 macrophages and in endotoxemic mice. J Immunol 157(10): p. 4634-40.
78.Le Moine, O., P. Stordeur, L. Schandene, A. Marchant, D. de Groote, M. Goldman, and J. Deviere. 1996. Adenosine enhances IL-10 secretion by human monocytes. J Immunol 156(11): p. 4408-14.
79.Hasko, G., Z.H. Nemeth, E.S. Vizi, A.L. Salzman, and C. Szabo. 1998. An agonist of adenosine A3 receptors decreases interleukin-12 and interferon-gamma production and prevents lethality in endotoxemic mice. Eur J Pharmacol 358(3): p. 261-8.
80.Link, A.A., T. Kino, J.A. Worth, J.L. McGuire, M.L. Crane, G.P. Chrousos, R.L. Wilder, and I.J. Elenkov. 2000. Ligand-activation of the adenosine A2a receptors inhibits IL-12 production by human monocytes. J Immunol 164(1): p. 436-42.
81.McKee, C.M., C.J. Lowenstein, M.R. Horton, J. Wu, C. Bao, B.Y. Chin, A.M. Choi, and P.W. Noble. 1997.Hyaluronan fragments induce nitric-oxide synthase in murine macrophages through a nuclear factor kappaB-dependent mechanism. J Biol Chem 272(12): p. 8013-8.
82.Horton, M.R., M.D. Burdick, R.M. Strieter, C. Bao, and P.W. Noble. 1998. Regulation of hyaluronan-induced chemokine gene expression by IL-10 and IFN-gamma in mouse

macrophages. J Immunol 160(6): p. 3023-30.

83.Charlab, R., E.D. Rowton, and J.M. Ribeiro. 2000. The salivary adenosine deaminase from the sand fly *Lutzomyia longipalpis*. Exp Parasitol 95(1): p. 45-53.

84.Hasko, G., D.G. Kuhel, Z.H. Nemeth, J.G. Mabley, R.F. Stachlewitz, L. Virag, Z. Lohinai, G.J. Southan, A.L. Salzman, and C. Szabo. 2000. Inosine inhibits inflammatory cytokineproduction by a posttranscriptional mechanism and protects against endotoxin induced shock. J Immunol 164(2): p. 1013-9.

85.Titus, R. and M. Mbow.1999. The vasodilator of *Lutzomyia longipalpis* sand fly salivary glands exacerbates infection with *Leishmania major* in mice. FASEB J 13(5): p. A970. Part 2 Suppl. S.

86.Kamhawi, S., Y. Belkaid, G. Modi, E. Rowton, and D. Sacks. 2000. Protection against cutaneous leishmaniasis resulting from bites of uninfected sand flies [In Process Citation]. Science 290(5495): p. 1351-4.

87.Valenzuela, J.G., Y. Belkaid, M.K. Garfield, S. Mendez, S. Kamhawi, E.D. Rowton, D.L. Sacks, and J.M.C. Ribeiro. 2001. Toward a defined anti-*Leishmania* vaccine targeting vector antigens: Characterization of a protective salivary protein. J. Exp. Med 194: p. 331 342.

MEMBRANE TRANSPORT AND METABOLISM IN *LEISHMANIA* PARASITES

Scott M. Landfear

Department of Molecular Microbiology and Immunology
Oregon Health Sciences University, Portland, OR 97201

INTRODUCTION

Membrane transporters that mediate the uptake of hydrophilic compounds across hydrophobic membrane bilayers are important components of metabolic pathways in all organisms. The advent of molecular genetic approaches has considerably advanced our understanding of such permeases and their functions in cellular metabolism. In this chapter, I summarize recent discoveries concerning four membrane transport systems that have been studied in some detail in *Leishmania* parasites, specifically those for glucose, GDP-mannose, biopterin, and nucleosides. Progress in the *Leishmania* genome project promises to make feasible in depth biochemical, mechanistic, genetic, and cell biological investigations of additional transporters whose genes have not yet been identified and analyzed in molecular detail. As in other organisms, a substantial proportion of the *Leishmania* genome is likely to be devoted to encoding membrane transport proteins.

MEMBRANE TRANSPORTERS

Transporters or permeases play a central role in the metabolism of *Leishmania* and other parasitic protozoa, as they mediate the uptake of nutrients from both the insect vectors and the vertebrate hosts in which the parasite resides during its entire life cycle. The essential function of these integral membrane proteins is to facilitate the migration of more or less hydrophilic solutes across the hydrophobic membrane bilayer. This activity is achieved by providing a substrate selective 'permeation pathway' that acts as an alternating access pore 1, allowing binding of the substrate on one side of the membrane followed by a conformational change that subsequently exposes the substrate to the opposite side of the membrane. Permeases are typically polytopic integral membrane proteins (Fig.1) that span the lipid bilayer multiple times *via* hydrophobic segments that often assume α-helical or sometimes β-sheet secondary structures. Hydrophobic faces of these transmembrane segments may allow stability of the structure in the lipid bilayer or interaction with other hydrophobic regions of the protein, but hydrophilic faces or patches on these segments may form components of the permeation pore that interacts with solutes or with the aqueous phase 2.

This chapter constitutes a brief review of some membrane transport systems from various *Leishmania* species. Given the space limitations, no attempt has been made to be exhaustive or inclusive of all the transporters that have been examined, and thus many interesting and worthy studies have not been covered. Rather, I have selected a few examples of permeases whose activities are specifically relevant to metabolism of these protozoa, whose genes have been cloned thus providing genetic insights into the uptake processes, and that illustrate some interesting aspects of transporter biology.

TRANSPORTERS

Glucose is a major source of metabolic energy for many eukaryotic cells and is thought to be especially important in the promastigote stage of the *Leishmania* life cycle. While parasites are initially ingested by the sandfly in a blood meal, the blood is subsequently digested, and the insects feed upon plant nectar that is rich in sugars including sucrose (3). This sugar, which is stored in the crop and released into the midgut, is subsequently cleaved to glucose and fructose, both of which can be taken up by the promastigotes and metabolized to yield energy. Glucose transport by whole parasites has been studied by various groups, revealing the presence of substrate saturable permeases that mediate the uptake of both glucose and fructose (4-7). In contrast, the amastigotes that reside within macrophage phagolysosomes have less ready access to free glucose, and both glucose transport (4) and activities of glycolytic enzymes (8-10) are down-regulated in amastigotes compared to promastigotes.

Glucose transporter genes have been cloned from several species of *Leishmania*. In original studies with *L. enriettii*, a family of clustered genes was identified (11,12) that encoded two isoforms, ISO1 and ISO2 (13). Both proteins are related in sequence and predicted secondary structure to the mammalian facilitative glucose transporters (14) and are thought to consist of 12 α-helical transmembrane segments connected by hydrophilic loops and to contain hydrophilic domains at the NH_2- and COOH-termini that are located on the cytoplasmic face of the plasma membrane (Fig.1).

These two isoforms are identical in sequence except for the NH_2-terminal hydrophilic domains that are completely divergent (13), and both ISO1 and ISO2 are functional glucose transporters when expressed in *Xenopus* oocytes (15). The most notable difference between these two permeases is that ISO1 is targeted largely to the flagellar membrane FM, whereas ISO2 is targeted to the pellicular plasma membrane PPM that

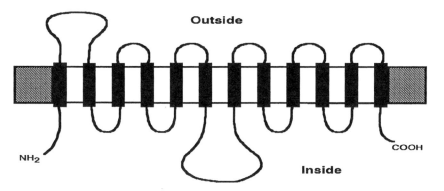

Figure 1. *Predicted topology of facilitative glucose transporter family members. The solid rectangles represent the 12 predicted transmembrane segments,* and *the loops are hydrophilic domains. The cross-hatched boxes represent the lipid bilayer.*

surrounds the cell body (16). The unique NH_2-terminal domain of ISO1 is responsible for flagellar targeting, as chimeras between this domain and the body of a PPM-resident transporter traffic to the FM, whereas chimeras containing the NH_2-terminal domain of ISO2 traffic to the PPM (17). Both *ISO1* and *ISO2* mRNAs are expressed primarily in promastigotes and are down-regulated in amastigotes (13). A related cluster of genes encompassing one copy of *ISO1* followed by four copies of *ISO2* has been identified in *L. donovani* (18).

Glucose transporter genes have also been studied in *L. mexicana*, where there is a cluster of three genes, *LmGT1*, *LmGT2*, and *LmGT3*, that encode distinct isoforms (19). The encoded proteins differ most notably in their NH_2-terminal domains, but there are also differences elsewhere in the sequence. One notable distinction compared to *L. enriettii* is that only one of these mRNAs, *LmGT2*, is strongly down-regulated in amastigotes, whereas *LmGT1* and *LmGT3* mRNAs are constitutively expressed during both stages of the life cycle. Studies now in progress have identified important differences between these three isoforms in transport properties, subcellular localization, and biological function within the intact parasite including an important role for glucose transport in the amastigote stage of the life cycle (R.J.S. Burchmore and S.M. Landfear, manuscript in preparation).

There has been a long-standing controversy concerning the mechanism of action of glucose transporters in these parasites. Thus, early studies based on sensitivity to treatments that collapse transmembrane potential or proton gradients or measurement of radiolabeled glucose accumulation inside the parasite (7) suggested that the glucose transporters may function as proton symporters (Fig.2).

Landfear

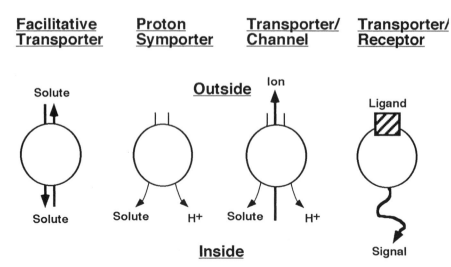

Figure 2. **Modes of Transporter Activity. Facilitative transporters** shuttle solute across the membrane but do not concentrate it on one side of the bilayer. **Proton symporters** couple the flux of solute to import of a proton, thus providing energy for concentration of the solute. Less conventional modes of activity include permeases that also function as **ion channels** and transporters or transporter-like proteins that act as signal transduction **receptors** by binding a ligand and transducing a signal to the interior of the cell.

Such an active transporter would concentrate glucose within the cell by harnessing the energy released when a proton is co-transported into the electronegative environment of the cytoplasm. However, sensitivity to metabolic inhibitors and concentration of radiolabel could be explained by the metabolic conversion of glucose to glucose-6-phosphate with the hydrolysis of ATP and the subsequent trapping of the phosphorylated glucose within the cell. Furthermore, transport measurements on *L. donovani* promastigotes grown in a chemostat (20) as well as failure to detect proton-mediated currents in oocytes expressing the ISO2 transporter (15) raised the possibility that *Leishmania* glucose transporters might function a facilitative permeases (Fig.2) that do not actively transport their substrates, much like their mammalian homologs. Notwithstanding, more recent studies with isolated membranes from promastigotes of *L. donovani* (21), which should not be complicated by the metabolism of glucose by cytosolic enzymes, support a proton-symport mechanism. Thus, application of a pH gradient to sealed membrane ghosts drove the accumulation of 2-deoxy-D-glucose inside the vesicles, and activation of a Mg^{+2}-ATPase that is an H^+/K^+-antiporter in everted membrane vesicles drove the extrusion of preloaded radiolabeled D-glucose from the vesicles. This likely active transport mechanism may promote import of glucose in the fluctuating glucose concentration present in the sandfly midgut.

Another permease that is related in structure and sequence to these glucose transporters is the *myo*-inositol transporter, MIT, of *L. donovani* (22). *myo*-inositol is a precursor in the biosynthesis of glycosyl-phosphatidylinositol lipids that constitute membrane anchors for many abundant surface proteins or glycolipids of *Leishmania* and related parasites (23), and this transporter presumably promotes the salvage of this precursor from the host. MIT has been identified as a proton symporter by electrophysiological characterization of currents detected in *Xenopus* oocytes expressing MIT (24). Furthermore, site-directed mutagenesis of MIT has identified two charged residues located within predicted transmembrane segments (25) and four conserved sequence motifs in intracellular hydrophilic loops (26) that are important for function of the permease.

Golgi-Localized Transporter for GDP-mannose.

Transporters located in membranes of intracellular organelles also play important roles in the metabolism of eukaryotic cells by mobilizing precursors or nutrients between membrane bound compartments. A notable contribution to the understanding of organellar permeases has been made by the identification and functional characterization of a GDP-mannose permease that is targeted to the membrane of the Golgi apparatus (27). The initial identification of this transporter relied heavily upon genetic approaches. A major virulence factor of *L. donovani* is the abundant surface glycoconjugate lipophosphoglycan (LPG) (28), which contains a backbone of galactose-mannose phosphate repeat units linked to a phosphatidylinositol membrane anchor. Synthesis of these repeat units occurs within the Golgi lumen and requires nucleotide sugar donors for both galactose and mannose-1-phosphate, and these nucleotide sugars are synthesized in the cytosol and must be transported across the Golgi membrane to their site of utilization. Initial genetic studies aimed at identifying components of the LPG assembly pathway generated a mutant, C3PO (28), that was deficient in LPG biosynthesis, but crude microsomal preparations from this mutant were still able to synthesize LPG, indicating that the biosynthetic enzymes were functional. Complementation of the C3PO mutant identified the *LPG2* gene as the genetic lesion, and sequencing of this gene predicted a polytopic hydrophobic membrane protein as the gene product. Subsequent studies showed that GDP-mannose could be taken up by purified microsomal vesicles from wild type parasites but not from the C3PO mutant and that transfection of the *LPG2* gene into the mutant could restore Golgi uptake of this nucleotide sugar. Furthermore, epitope tagged LPG2 protein localized to the Golgi membrane, as determined by electron microscopy. Together these results strongly support the identification of LPG as a Golgi associated GDP-mannose transporter that is essential for delivery of this substrate for synthesis of the virulence factor LPG.

Notably, mammalian cells do not contain a GDP-mannose transporter, thus raising the possibility that the LPG2 permease could be a target for chemotherapy. Further characterization of this interesting permease might lead to identification of transport inhibitors that could be lead compounds for drug development. This protein is however related in sequence to a number of predicted proteins from various prokaryotes and eukaryotes, several of which are nucleotide sugar transporters including the yeast Golgi GDP-mannose transporter. Biochemical studies on these transporters have revealed that they are antiporters that exchange lumenal nucleoside monophosphate for the nucleotide sugar, thus concentrating the latter substrate inside the Golgi (29). It is virtually certain that *Leishmania* parasites also express other nucleotide sugar transporters that promote the uptake of different nucleotide sugars into the Golgi.

BIOPTERIN TRANSPORTER

Leishmania and related parasites require pteridines folates and pterins for growth, but unlike their vertebrate hosts, they cannot synthesize pterins *de novo* (30). Biopterin and other unconjugated pterins can promote growth of *Leishmania* in folate-deficient medium, and biopterin is an essential nutrient (31). The nutritional importance of pteridines has focused interest on their modes of transport into these parasites. A gene encoding a biopterin transporter, designated either *BT1* or *ORFG*, has been cloned from both *L. donovani* (32) and *L. tarentolae* (30) and has lead to the functional characterization of this metabolically important permease.

Initial studies in *L. donovani* revealed the presence of an amplified extrachromosomal circular DNA, designated LD1, or a related linear minichromosome, in ~15% of the strains tested (32). One of the open reading frames (ORFs) in LD1 encoded a protein with 12 predicted transmembrane domains, similar to many transporter proteins (33). Subsequent studies showed that plasmids expressing *ORFG* could rescue a methotrexate-resistant folate transport-deficient mutant of *L. donovani*, and that these transfectants exhibited increased uptake of both biopterin and folate, suggesting that *ORFG* encoded a biopterin/folate transporter (32). Furthermore, expression of *ORFG* in *Xenopus* oocytes confirmed its biopterin transport activity, and the gene was consequently renamed *BT1*. A *BT1* null mutant, generated by targeted gene disruption, was deficient in biopterin uptake, but biopterin transport was restored to this knockout line by expression of the *BT1* gene on an extrachromosomal element. It is notable that the *BT1* null mutants could not be generated unless the medium was supplemented with biopterin and folate and that the knockout line grew more slowly than the wild type even in the presence of biopterin and folate. These results suggest that pterins are essential for growth and that they may enter the parasite by passive diffusion or by another low affinity transporter in the absence of functional BT1 transporter. Furthermore, the importance of biopterin uptake for parasite growth suggests that the amplification of the *BT1* gene on LD1 or other amplified DNA elements may confer a growth advantage on such strains, accounting for the frequency of such amplifications seen in *Leishmania* isolates. Biopterin plays another undefined role in parasite metabolism, and the parasite's inability to synthesize pterins thus makes uptake of host biopterin critical for *Leishmania* survival. Finally, substrate specificity studies showed that BT1 is a high affinity biopterin transporter and a low affinity folate transporter, whereas another permease promotes the high affinity uptake of folate.

In separate studies (30) the *ORFG* gene was functionally cloned from *L. tarentolae* by its ability to confer methotrexate resistance upon the parasite. Growth in the presence of methotrexate selected for loss of the high affinity folate transporter, and complementation with extrachromosomal elements that overexpress *ORFG* apparently rescued the folate deficiency by inducing folate uptake and/or by promoting synthesis of folate from biopterin. Similar to the studies described above, knockout of the *ORFG* gene resulted in biopterin transport deficiency, and complementation of the null line with the *ORFG* gene restored biopterin transport. Epitope tagged ORFG protein was also shown to accumulate in the plasma membrane surrounding the cell body.

NUCLEOSIDE TRANSPORTERS

Another family of transporters that is of considerable interest from both the metabolic and pharmacological point of view includes those that mediate the uptake of nucleosides. All parasitic protozoa studied to date are unable to synthesize purines *de novo* and therefore rely upon uptake and salvage of preformed purines for survival (34). These purines can be salvaged as either nucleosides or nucleobases, and the permeases that promote the uptake of these two classes of purines appear to be different. Furthermore, some purine analogs of the pyrazolopyrimidine family, including formycin B and allopurinol riboside, are selectively cytotoxic to the parasite, as they are recognized by the parasite salvage enzymes and are ultimately incorporated into parasite RNA, whereas these analogs are not significantly metabolized by the mammalian salvage enzymes (35). These cytotoxic compounds are taken up by the parasite nucleoside transporters, which are thus essential for drug delivery,

Biochemical and genetic studies on *L. donovani* established that there are two distinct nucleoside transporters, one of which mediates the uptake of adenosine and the pyrimidine nucleosides (now referred to as LdNT1) and the other of which transports inosine and guanosine (LdNT2) (36). Lines that were genetically deficient in LdNT1 activity were isolated by mutagenesis followed by selection in the presence of the cytotoxic adenosine analog tubercidin (TUBA5 cell line), and another line deficient in LdNT2 activity was selected with the inosine analog formycin B (FBD5 line). These drug-resistant mutant lines were subsequently used to clone the *LdNT1* (37) and *LdNT2* (38) genes, by transfecting each line with a cosmid genomic library and screening for clonal transfectants that regained sensitivity to tubercidin or formycin B respectively.

Molecular analysis of one of the cosmids so identified revealed that the *LdNT1* locus consisted of two ORFs, designated *LdNT1.1* and *LdNT1.2*, that differed from each other in 6 codons. Each ORF encoded a protein with significant sequence identity to the mammalian equilibrative nucleoside transporters (ENTs) (39) that are broadly distributed among different tissues and that mediate the non-concentrative uptake of nucleosides. Both the *LdNT1.1* and *LdNT1.2* ORFs were capable of conferring tubercidin sensitivity and restoring adenosine and pyrimidine nucleoside transport capacity when transfected into TUBA5 mutants, and they also induced adenosine and uridine transport when expressed in *Xenopus* oocytes. *LdNT1.1* but not *LdNT1.2* mRNA was detected on Northern blots of promastigote RNA, suggesting that the latter gene is expressed at very low levels. Also of interest, electrophysiological studies on these permeases revealed the presence of substrate-induced currents that increased with decreasing pH (M. Kavanaugh *et al.*, manuscript in preparation), strongly suggesting that they are active proton symporters that can concentrate nucleosides within the parasite, a property that would be useful for the salvage of critical nutrients.

Further studies to investigate the nature of the mutations that resulted in transport deficiency in the TUBA5 line (40) revealed the presence of single point mutations in the two alleles of the LdNT1.1 gene. The mutation in one allele caused a G183D change, located on the hydrophilic face of the predicted amphipathic transmembrane helix number 5 (TM5), and the mutation in the other allele resulted in a C337Y alteration within a hydrophilic patch of predicted transmembrane helix 7 (TM7). Site-directed mutagenesis of the wild type LdNT1.1 confirmed that each point mutation resulted in loss of almost all of the adenosine transport activity, but green fluorescent protein (GFP) fusions of each of the mutant transporters trafficked efficiently to the plasma membrane, indicating that each mutation disrupted transport activity *per se* and not delivery of the permease to the surface membrane. One potential explanation for the transport-deficient phenotype of each mutant is that amino acids 183 and 337 lie on the permeation pathway of the transporter, consistent with their location in hydrophilic faces of predicted TM helices, and that alteration of each amino acid prevents the binding of substrate or its movement across the pathway. In this regard, it is especially interesting that a 'conservative' G183A mutant retained substantial adenosine transport capacity but was almost completely deficient in uridine transport. The observation that this mutant had undergone a change in substrate specificity is consistent with G183 being part of or near a substrate binding site and further suggests that TM5 constitutes part of the permeation pore. Further biochemical and genetic studies should help to map the regions of LdNT1.1 and related transporters that serve the essential function of substrate permeation across the membrane.

Similar studies on *LdNT2* (38) have revealed that it is a single copy gene whose ORF is ~33% identical in amino acid sequence to LdNT1.1 and is also related to the mammalian ENT family of transporters. Expression of *LdNT2* in the transport-deficient FBD5 line or in *Xenopus* oocytes restored uptake of inosine and guanosine, confirming that it is a functional purine nucleoside transporter. Alignment of the LdNT2 sequence with other functional nucleoside transporters from mammals and protozoa demonstrated the existence of 16 conserved residues in a sequence of 499 amino acids (41). Some of these residues may represent amino acids that are of particular importance for folding or transport function of this family of related transporters.

Recently, a study on *L. donovani* amastigotes isolated from infected hamster spleens uncovered a novel amastigote-specific purine nucleoside transport activity designated T_2 (42). This permease differed in substrate specificity, recognizing adenosine, guanosine, and inosine, and developmental regulation from LdNT1.1 and LdNT2, but it was similar to these transporters in being a high affinity permease with a K_m for adenosine of ~2 μM. To date, the gene for this transporter has not been cloned and characterized. It will be interesting to determine whether there are functional differences of the T_2 permease that are adaptive for activity within the macrophage phagolysosome.

CONCLUSIONS

The observation that as many as 5% of the genes in *Saccharomyces cerevisiae* may be devoted to transport activities (43) suggests that there are likely to be numerous transporters in *Leishmania* parasites that serve either essential or supportive functions for survival of these parasites within their vertebrate and invertebrate hosts. Although I have touched on only a few such activities here, rapid advances in identification and characterization of other permeases are likely to be the standard as the genome project of *L. major* reaches completion (44). Genomic approaches are of particular utility with regard to this class of polypeptides, as the difficulties inherent in purification of non-abundant membrane proteins have been an obstacle to the classical biochemistry of transporters.

Over the past decade, advances in the molecular genetics of *Leishmania* parasites, such as the ability to clone, sequence, and functionally express genes, to perform classical or site directed mutagenesis, to localize transporters by immunofluorescence, epitope tagging, or GFP fusions, to clone transporter genes by functional complementation, and to critically test biological functions by targeted gene replacement followed by complementation have provided insights that were difficult to attain prior to these technical advances. In addition, biochemical analysis of intact cells provides a global image of transport function but is also complicated by the fact that most such transport is probably the aggregate activity of multiple permeases. The functional expression of distinct genes now provides the opportunity to dissect individual transporters and to understand how they differ in biologically significant ways from related family members. We should thus be able to achieve a global picture of each cellular transport activity by characterization of its component parts. Of particular interest is the comparison of transporters that are expressed at different developmental stages and thus function within distinct physiological environments and of permeases that are localized to different compartments of these unicellular organisms and may thus subserve unique biological functions.

In recent years, several unconventional activities have been identified for transporters in various organisms. Thus some permeases can function both as alternating access pores and as channels (1) that pass large numbers of ions in an open channel mode (Fig.2). In addition, other transporters or transporter-like proteins can function as signal transduction receptors that bind ligand and transmit a signal to the interior of the cell (Fig.2). Thus, two glucose transporter-like proteins in *S. cerevisiae* are glucose sensors that are involved in regulation of expression of genes for *bona fide* glucose transporters (45), and several other permeases in organisms as diverse as bacteria and humans serve well-established signal transduction functions, often in addition to conventional transport activities (46, 47). As we study these intriguing proteins within *Leishmania* and related parasites, we should be alert for the possibility of these or other unconventional functions. Thus, we have suggested that the flagellar glucose transporter of *L. enriettii* ISO1, and its homolog in *L. mexicana* LmGT1, might serve as glucose sensors (48). This speculation is prompted by the observation that flagellar or ciliary membrane proteins in other eukaryotes are often involved in signal transduction processes such as neuronal sensing or mating type response, and that several other flagellar-specific proteins in trypanosomes have structural homologies with well characterized components of signal transduction pathways in higher eukaryotes (48). Furthermore, the ability of *Leishmania* promastigotes to chemotax toward glucose and fructose (49) provides a potential requirement for glucose sensing in these protozoa. Future studies on these and other transporters will be required to determine whether or not they are involved in sensing the extracellular environment.

ACKNOWLEDGEMENTS

Some of the work reported in this chapter was supported by grants AI25920 and AI44138 from the National Institutes of Health. S.M.L. is a Burroughs Wellcome Molecular Parasitology Scholar. I would also like to thank Dr. Buddy Ullman for critical comments on the manuscript.

REFERENCES

1. Kavanaugh, M. P. 1998. Neurotransmitter transport: models in flux Proc. Natl Acad Sci U.S.A. 95:12737-12738.

2. Doyle, D. A., Cabral, J. M., Pfeutzner, R. A., Kuo, A., Bulbis, J. M., Cohen, S. L., Chait,B. T. and MacKinnon, R. 1998. The structure of the potassium channel: molecular basis of K+ conduction and selectivity 280:69-78.

3. Schlein, Y. 1986. Sandfly diet and *Leishmania*. Parasitol Today 2: 175-177.

4. Burchmore, R. J. and Hart, D. T. 1995. Glucose transport in amastigotes and promastigotes of *Leishmania mexicana mexicana*. Mol Biochem Parasitol 74: 77-86.

5.. Schaefer, F. W. and Mukkada, A. J. 1976. Specificity of the glucose transport system in *Leishmania tropica* promastigotes J Protozool 23:446-449.

6. .Zilberstein, D. and Dwyer, D. 1984. Glucose transport in *Leishmania donovani* promastigotes. Mol Biochem Parasitol 12:327-336.

7. .Zilberstein, D. and Dwyer, D. 1985. Antidepressants cause lethal disruption of membrane function in the human protozoan parasite *Leishmania*. Proc Natl Acad Sci U.S.A. 82: 1716-1720.

8. Coombs, G. H., Craft, J. A. and Hart, D. T. 1982. A comparative study of *Leishmania mexicana* amastigotes and promastigotes. Enzyme activities and subcellular locations. Mol Biochem Parasitol 5: 199-211.

9. Hart, D. T. and Coombs, G. H. 1982. *Leishmania mexicana*: energy metabolism of amastigotes and promastigotes. Exp Parasitol 54: 397-409.

10. Rainey, P. M. and MacKenzie, N. E. 1991. A carbon-13 nuclear magnetic resonance analysis of the products of glucose metabolism in *Leishmania pifanoi* amastigotes and promastigotes. Mol Biochem Parasitol 45: 307-316.

11. Cairns, B. R., Collard, M. W. and Landfear, S. M. 1989. Developmentally regulated gene from *Leishmania* encodes a putative membrane transport protein. Proc Natl Acad Sci U.S.A. 85:2130-2134.

12. Stein, D. R., Cairns, B. R. and Landfear, S. M. 1990. Developmentally regulated transporter in *Leishmania* is encoded by a family of clustered genes. Nucleic Acids Res 18: 1549-1547.

13. Stack, S. P., Stein, D. R. and Landfear, S. M. 1990. Structural isoforms of a membrane transport protein from *Leishmania enriettii*. Mol Cell Biol 10: 6785-6790.

14. Silverman, M. 1991. Annu Rev Biochem 60: 757-794.

15. Langford, C. K., Little, B. M., Kavanaugh, M. P. and Landfear, S. M. 1994. Functional expression of two glucose transporter isoforms from the parasitic protozoan *Leishmania enriettii*. J Biol Chem 269: 17939-17943.

16. Piper, R. C., Xu, X., Russell, D. G., Little, B. M. and Landfear, S. M. 1995. Differential targeting of two glucose transporters from *Leishmania enriettii* is mediated by an NH2-terminal domain. J Cell Biol 128:499-508.

17. Snapp, E. L. and Landfear, S. M. 1997. Cytoskeletal association is important for differential targeting of glucose transporter isoforms in *Leishmania*. J Cell Biol 139:1775-1783.

18. Bringaud, F., Vedrenne, C., Cuvillier, A., Parzy, D., Baltz, D., Tetaud, E., Pays, E., Venegas, J., Merlin, G. and Baltz, T. 1998. Conservation of metacyclic variant surface glycoprotein expression sites among different trypanosome isolates. Mol Biochem Parasitol 94: 249-264.

19. Burchmore, R. J. S. and Landfear, S. M. 1998. Differential regulation of multipleglucose transporter genes in *Leishmania mexicana*. J Biol Chem 273: 29118-29126.

20. ter Kuile, B. H. and Opperdoes, F. R. 1993. Uptake and turnover of glucose in *Leishmania donovani*. Mol Biochem Parasitol 60: 313-322.

21. Mukherjee, T., Mandal, D. and Bhaduri, A. 2001. Leishmania plasma membrane Mg2+-ATPase is a H+/K+-antiporter involved in glucose symport. Studies with sealed ghosts and vesicles of opposite polarity. J Biol Chem 276: 5563-5569.

22. Drew, M. E., Langford, C. K., Klamo, E. M., Russell, D. G., Kavanaugh, M. P. and Landfear, S. M. 1995. Functional expression of a myo-inositol/H+ symporter from *Leishmania donovani*. Mol Cell Biol 15: 5508-5515.

23. Werbovetz, K. A. and Englund, P. T. 1997. Glycosyl phosphatidylinositol myristoylation in African trypanosomes. Mol Biochem Parasitol 85: 1-7.

24. Klamo, E. M., Drew, M. E., Landfear, S. M. and Kavanaugh, M. P. 1996. Kinetics and stoichiometry of a proton/myo-inositol cotransporter. J Biol Chem 271: 14937-14943.

25. Seyfang, A., Kavanaugh, M. P. and Landfear, S. M. 1997. Aspartate 19 and glutamate 121 are critical for transport function of the myo-inositol/H+ symporter from *Leishmania donovani*. J Biol Chem 272: 24210-24215.

26. Seyfang, A. and Landfear, S. M. 2000. Four conserved cytoplasmic sequence motifs are important for transport function of the *Leishmania* inositol/H+ symporter. J Biol Chem 275: 5687-5693.

27. Ma, D., Russell, D. G., Beverley, S. M. and Turco, S. J. 1997. Golgi GDP-mannose uptake requires *Leishmania* LPG2. A member of a eukaryotic family of putative nucleotide-sugar transporters. J Biol Chem 272: 3799-3805.

28. Descoteaux, A., Luo, Y., Turco, S. A. and Beverley, S. M. 1995. A specialized pathway affecting virulence glycoconjugates of *Leishmania*. Science 269: 18691872.

29. Abeijon, C., Mandon, E. C. and Hirschberg, C. B. 1997. Trends Biochem Sci 22: 203-207.

30. Kündig, C., Haimeur, A., Legare, D., Papadopoulou, B. and Ouellette, M. 1999. Increased transport of pteridines compensates for mutations in the high affinity folate transporter and contributes to methotrexate resistance in the protozoan parasite *Leishmania tarentolae*. EMBO J 18: 2342-2351.

31. Beck, J. T. and Ullman, B. 1991. Biopterin conversion to reduced folates by *Leishmania donovani* promastigotes. Mol Biochem Parasitol 49: 21-28.

32. Lemley, C., Yan, S., Dole, V., Madhubala, R., M.L., C., Beverley, S. M., Myler, P. J. and Stuart, K. D. 1999. The *Leishmania donovani* LD1 locus gene ORFG encodes a biopterin transporter BT1. Mol Biochem Parasitol 104: 93-105.

33. Pao SS, Paulsen IT, Saier MH Jr. 1998. Major facilitator superfamily. Microbiol Mol Biol Rev 62: 1-34.

34. Berens, R. L., Krug, E. C. and Marr, J. J. 1995. In Biochemistry and Molecular Biology of Parasites. eds. Marr, J. J. and Müller, M. Academic Press, New York 89-117.

35. Ullman, B. 1984. Pharmaceut Res 1: 194-203.

36. Aronow, B., Kaur, K., McCartan, K. and Ullman, B. 1987. Two high affinity nucleoside transporters in *Leishmania donovani*. Mol Biochem Parasitol 22: 29-37.

37. Vasudevan, G., Carter, N. S., Drew, M. E., Beverley, S. M., Sanchez, M. A., Seyfang, A., Ullman, B. and Landfear, S. M. 1998. Cloning of *Leishmania* nucleoside transporter genes by rescue of a transport-deficient mutant. Proc Natl Acad Sci U.S.A. 95: 9873-9878.

38. Carter, N. S., Drew, M. E., Sanchez, M., Vasudevan, G., Landfear, S. M. and Ullman, B. 2000. Cloning of a novel inosine-guanosine transporter gene from *Leishmania donovani* by functional rescue of a transport-deficient mutant. J Biol Chem 275: 20935-20941.

39. Hyde, R. J., Cass, C. E., Young, J. D. and Baldwin, S. A. 2001. The ENT family of eukaryote nucleoside and nucleobase transporters: recent advances in the investigation of structure/function. Mol Membrane Biol 18: 53-63

40. Vasudevan, G., Ullman, B. and Landfear, S. M. 2001. Point mutations in a nucleoside transporter gene from *Leishmania donovani* confer drug resistance and alter substrate selectivity. Proc Natl Acad Sci U. S. A. 98: 6092-6097.

41. Carter, N. S., Landfear, S. M. and Ullman, B. 2001. Nucleoside transporters of parasitic protozoa. Trends Parasitol 17: 142-145.

42. Ghosh, M. and Mukherjee, T. 2000. Mol Biochem Parasitol 108: 93-99.

43. Goffeau, A., Barrell, B. G., Bussey, H., Davis, R. W., Dujon, B., Feldmann, H., Gailbert, F., Hoheisel, J. D., Jacq, C., Johnston, M., Louis, E. J., Mewes, H. W., Murakami, Y., Phillippsen, P., Tettelin, H. and Oliver, S. G. 1996. Life with 6000 genes. Science 274: 562-567.

44. Myler, P. J. and Stuart, K. D. 2000. Curr Opin Microbiol 3: 412-416.

45. Özcan, S., Dover, J., Rosenwald, A. G., Wölf, S. and Johnston, M. 1996. Two glucose transporters in *Saccharomyces cerevisiae* are glucose sensors that generate a signal for induction of gene expression. Proc Natl Acad Sci U.S.A. 93: 12428-12432.

46. Bandyopadhyay, G., Sajan, M. P., Kanoh, Y., Standaert, M. L., Burke, T. R., Quon, M. J., Reed, B. C., Dikic, I., Noel, L. E., Newgard, C. B. and Farese, R. 2000 Glucose activates mitogen-activated protein kinase extracellular signal-regulated kinase through proline-rich tyrosine kinase-2 and the Glut1 glucose transporter J. Biol. Chem. 275: 40817-40826.

47. Rolland, F., Winderickx, J. and Thevelein, J. M. 2001 Glucose-sensing mechanisms in eukaryotic cells Trends Biochem. Sci. 26: 310-317.

48. Landfear, S. M. and Ignatushchenko, M. 2001 The flagellum and flagellar pocket of trypanosomatids. Mol. Biochem. Parasitol. 115, 1-17.

49. Oliveira, J. S., Melo, M. N. and Gontijo, N. F. 2000 A sensitive method for assaying chemotaxic responses of *Leishmania* promastigotes. Exp. Parasitol. 96:187-189.

THE INTERACTION OF *LEISHMANIA* SPP. WITH PHAGOCYTIC RECEPTORS ON MACROPHAGES: THE ROLE OF SERUM OPSONINS

David M. Mosser[1] and Andrew Brittingham[2]

[1]Cell Biology and Molecular Genetics, University of Maryland
College Park, MD 20742
[2]Department of Microbiology, Des Moines University
Osteopathic Medical Center, Des Moines, IA 50125

INTRODUCTION

There are two developmental forms of Leishmania; the promastigote and the amastigote. In this chapter we describe the interaction of each of these developmental forms with host macrophages, and we discuss how these molecular interactions may influence subsequent parasite development. We present evidence that each of the two developmental forms exploits a different specific host opsonin to facilitate their entry into and survival within host macrophages. The promastigote form utilizes host complement component C3, whereas the amastigote form uses host IgG.

PROMASTIGOTES

Following their inoculation into a vertebrate host, during sandfly feeding, *Leishmania* promastigotes immediately encounter host serum, and the lytic and opsonic factors that it contains. The ability of *Leishmania* promastigotes to resist serum lysis is dependent on the expression of gp63 (1), and this resistance has been reviewed previously (2). The fixation of complement and the exploitation of this opsonin will be the focus of this review. An early examination of the mechanisms of complement activation by promastigotes demonstrated that both *L.major* and *L.enrietti* were capable of activating complement via the alternative pathway, a process that proceeded in the absence of antibody (3). The activation of the alternative complement pathway was later extended to include *L.mexicana*, *L.amazonensis*, and *L.braziliensis* (4). Complement activation by *L.donovani* (5) and also metacyclic *L.major* promastigotes (6) also involves the classical complement pathway. Several reports have suggested the involvement of naturally occurring antibodies in the classical activation of complement by *L.donovani* (4,5,7) and *L.amazonensis* (7). Other reports have demonstrated the binding of acute phase proteins, including C-reactive protein (CRP) (8) and mannan binding proteins (MBP) (9) to promastigotes. CRP and MBP have been demonstrated to be capable of activating a complement through a lectin pathway involving components of the classical pathway (10,11). Thus the parasite appears to have developed multiple redundant ways to assure itself of being opsonized by complement following its interaction with host serum.

Regardless of the mechanism of complement activation, deposition of C3 onto the surface of promastigotes has been shown to dramatically enhance the interaction of the promastigote with mononuclear phagocytes. This is especially true when the assays are performed with human monocyte-derived macrophages, which bind poorly to promastigotes in the absence of exogenous complement (12).   Surface bound C3 (and it's cleavage products) can act as ligands for macrophage receptors, thereby mediating the attachment of promastigotes to macrophages. The receptors involved in the binding and phagocytosis of complement-opsonized promastigotes are the macrophage receptors for complement protein C3; Mac-1 (CD11b/ CD18), the receptor for iC3b, and CR1 (CD35) the receptor for C3b.

Both C3b and iC3b forms have been reported on the surface of promastigotes following serum opsonization (6, 13, 14, 15, 16), and the two major surface structures of promastigotes, LPG and gp63, have both been reported to efficiently induce complement activation. Gp63 has additionally been reported to facilitate the conversion of surface bound C3b to iC3b, simultaneously preventing progression of the complement cascade and formation of the lytic membrane attack complex, while generating the ligand for Mac-1 (1). The interaction of serum-opsonized (C3b) promastigotes with CR1 present on human erythrocytes has also been reported to occur, and this process is thought to facilitate the eventual interaction of serum-opsonized promastigotes with complement receptors on phagocytic cells (7).

Initial studies examining the interaction of complement- opsonized promastigotes with macrophages were performed with murine macrophages, which express Mac-1 but not CR1.   These studies demonstrated the importance of Mac-1 in promastigote adhesion (15, 16).   Studies using knockout mice demonstrated that while macrophages from CD18 -/- mice were able to bind serum-opsonized promastigotes, their phagocytosis rate was reduced by 50% relative to wild-type macrophages (17).   Working with human macrophages, Da Silva and colleagues demonstrated a role for CR1 in the binding of metacyclic *L. major* promastigotes to human macrophages (14).   Rosenthal et al. (18) demonstrated that while both procyclic and metacyclic *L. major* promastigotes can bind to both CR1 and Mac-1, the stable adhesion of complement-opsonized *Leishmania* is mediated by Mac-1. Antibodies to Mac-1, but not CR1, inhibited the phagocytosis of serum opsonized metacyclic *L. major* promastigotes (18). Thus, the complement-dependent adhesion of *Leishmania* to human macrophages is mediated by both Mac-1 and CR1, but Mac-1 is the primary receptor involved in the complement-dependent phagocytosis of *Leishmania* promastigotes.

Antibodies that block Mac-1 functions have also been reported to inhibit the binding of non-serum opsonized promastigotes to macrophages (15, 16). One explanation for this observation has been proposed by Blackwell et al., who have shown the presence of C3 on the surface of *L.donovani* promastigotes following their incubation with macrophages, under serum free conditions (19). Macrophages can synthesize and secrete all the components of the alternative complement pathway necessary for C3 fixation (20). Furthermore, these macrophage-secreted components have been suggested to mediate the adhesion of zymosan particles to Mac-1 (21). This has led to the suggestion that the phenomena of "local opsonization" may be involved in the adhesion of promastigotes to macrophage Mac-1.

In addition to enhancing promastigote adhesion to macrophages complement opsonization was also shown to exert an influence on the intracellular fate of parasites following phagocytosis. Studies using unselected stationary phase *L.major* promastigotes demonstrated that in the absence of opsonic complement, more than 95% of the promastigotes taken up by macrophages were killed. However, when parasites were ingested following opsonization with complement, there was a dramatic enhancement in intracellular survival (22). These results have been confirmed using a metacyclic population of *L.major*, which also exhibited a marked enhancement in intracellular survival following opsonization with serum complement (23). These results indicated that the survival of *L.major* in macrophages ws potentiated by complement.

Because of the mechanism of sandfly feeding, and the regurgitation of promastigotes into a blood pool, promastigotes presumably encounter macrophages in a serum-opsonized state. However, this does not negate the importance of the (opsonin-independent) direct binding of promastigotes to macrophages. In fact in vitro studies by several groups have suggested that efficient attachment and internalization of parasites by macrophages is likely dependent on multiple receptor-ligand interactions, including species-specific lectin adhesion. Early studies demonstrated that the attachment of non-opsonized *L.major* promastigotes to mouse macrophages could be partially inhibited by lactose and galactose (24), suggesting that a parasite glycoconjugate may be involved in the direct attachment of parasites to macrophages. The adhesion of *L.donovani* promastigotes to murine or human macrophages could be inhibited by mannan and other inhibitors of the mannose receptor (16, 25, 26). The direct binding of *L.major* promastigotes to murine macrophages has been suggested to involve another carbohydrate receptor, a β-glucan receptor.   The binding of *L.major* promastigotes was inhibited by the β -glucan containing compounds, laminarin and zymocel, while mannan was demonstrated to have no affect on promastigote binding (27).   While the identity of the β -glucan receptor involved in promastigote binding is unclear, Mac-1 has been identified as at least one of the β -glucan receptor expressed by mononuclear phagocytes (28). An alternative carbohydrate receptor implicated in the attachment of promastigotes to macrophages is the receptor for advanced glycosylation end products (AGE) (29).   AGE's arise from the time- dependent, non-enzymatic reaction of glucose and proteins (30).   The AGE receptors on macrophages may be involved in the removal of senescent proteins and cells (31).

The identification of an abundant glycoconjugate (LPG) on the surface of promastigotes (32, 33) initiated experiments to define a role for this molecule in the direct attachment of promastigotes to macrophages. Handman and Goding (34) demonstrated that affinity purified LPG from *L.major* bound specifically to murine macrophages.   Promastigote binding could also be blocked by incubating the promastigotes with F (ab) fragments of antibodies against LPG (34).   Further analysis of *L.major* LPG structure demonstrated that the PO_4-6 [Gal (β 1-3) Gal (β 1-3) Gal (β 1-3)] Gal (β1-4) Man α1- region of LPG was responsible for its binding to macrophages, and inhibited *L.major* promastigote binding (35).   This region of LPG is unique to *L.major* and not found in the LPG of other species of *Leishmania*. The specificity of this interaction was demonstrated by the fact that this phosphooligisaccharide repeat was capable of inhibiting *L.major* binding, but had no affect on *L.donovani* binding (35). The macrophage receptor that recognizes this region of LPG remains undetermined.   In light of recent studies by Ilg (36) demonstrating that the genetic disruption of LPG biosynthesis had no effect on the binding of *L.mexicana* promastigotes to macrophages, LPG's role in the promastigote-macrophage interaction warrants new investigation.

Like LPG, the other major surface molecule on promastigotes, gp63, has also been implicated in promastigote adhesion to macrophages. Purified protein, as well as antibodies against gp63 have both been shown to inhibit promastigote adhesion (37, 38) and gp63 coated particles have been demonstrated to bind to macrophages (38).   Restoration of gp63 expression on a gp63 deficient variant of *L.amazonensis* was shown to restore promastigote binding to murine as well as human macrophages, and an antisense-based approach used by Chen et al. (39) recently demonstrated that the disruption of gp63 expression in *L.amazonensis* not only alters the kinetics of macrophage binding, but also intracellular survival.

The mechanisms of gp63 mediated adhesion to macrophages, as well as the identity of macrophage receptors involved in gp63 recognition remains somewhat controversial. Original reports suggested that gp63 contained the amino acid sequence Arg-Gly-Asp (RGD), a sequence that is recognized by many receptors of the integrin family. Russell and Wright reported that this RGD-containing region of gp63 is recognized by the macrophage complement receptor type 3 (CR3, Mac-1, CD11b/CD18) (40). Further analysis, however, demonstrated that gp63 does not possess an RGD region (41). The lack of involvement of an RGD sequence in gp63 mediated attachment to Mac-1 was further supported by studies of Taniguchi, who demonstrated that the RGD region of complement protein C3, the normal ligand for Mac-1, was not necessary for receptor binding (42). Viable promastigotes, of several species failed to exhibit direct binding to fibroblasts that were transfected with constructs encoding Mac-1 (12), nor did they bind to substrates coated with purified Mac-1 (1, 12, 18, 43). Additionally Chinese hamster ovary cells (CHO) stably expressing gp63 on their surface exhibited no direct binding to purified Mac-1 coated substrates (1). A second set of receptors that have been implicated in the gp63-mediated attachment of promastigotes to macrophages are the cellular receptors for fibronectin. Antibodies against fibronectin have been shown to cross react with gp63 (44, 45), and to also inhibit the binding of promastigotes to macrophages (44, 46). Soteriadou and colleagues (45) demonstrated that an SRYD region of gp63 was antigenically similar to the RGDS region of fibronectin. Additionally, Rizvi reported the increased attachment and spreading of fibroblasts to surfaces coated with *Leishmania* antigen preparations rich in gp63 (44). These data suggest that receptors for fibronectin present on macrophages may be capable of recognizing a "fibronectin like" region of gp63 and mediate promastigote attachment to macrophages.

Two of the most abundant receptors for fibronectin are members of the β 1 integrin family, VLA-4 (Cd49d/CD29) and VLA-5 (CD49e/CD29). VLA-5 recognizes the Arg-Gly-Asp-Ser (RGDS) region of fibronectin (47). VLA-4, in contrast recognizes the CS-1 domain of fibronectin, which contains the amino acid sequence Glu-Ile-Leu-Asp-Val (EILDV) (48). VLA-4 and VLA-5 are both expressed on a wide range of tissues and cell types, with VLA-4 appearing primarily on hematopoietic cells and being prominently expressed on macrophages and activated lymphocytes (43). The cooperation between fibronectin and complement receptors has been suggested by several groups, who have observed that the presence of fibronectin can enhance the phagocytosis of complement opsonized particles (49, 50, 51).

We demonstrated that gp63 on *Leishmania* can bind specifically to human α4/β1 fibronectin receptors (43). This was the first identification of a specific fibronectin receptor with which *Leishmania* can interact. We also demonstrated that the SRYD region of gp63, a domain previously implicated in cell adhesion (40, 45), was not required for this interaction. Finally, although gp63 bound directly to α4/β1 receptors, the binding of parasites to these receptors was not very efficient. Furthermore parasite binding to human macrophages was minimal in the absence of complement, underscoring the inefficiency with which fibronectin alone functions as an adhesive mechanism for promastigotes. In the presence of complement, however, the interaction of gp63 with fibronectin receptors may cooperate with complement receptors to mediate the efficient internalization of parasites by macrophages. Blocking fibronectin receptors delayed the internalization of complement-opsonized parasites into macrophages (43).

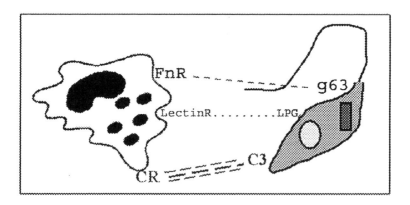

Figure 1. *The binding of promastigotes to macrophages may involve the interaction of several receptors on macrophages with their respective parasite ligands.*

AMASTIGOTES

In comparison to the promastigote form, much less is known about the interaction of amastigotes with macrophages. There are several reasons for this, including the fact that these organisms are more difficult to work with, since only a few species can be reliably grown in cell-free culture (52). Other potential complications in working with this form are the early observations that these organisms sometimes have cellular debris and a variety of serum proteins adsorbed to them (53). These and a variety of other factors may contribute to the lack of information and the lack of support for studies on amastigote infectivity. A review of the NIH CRISP database shows that the vast majority of active proposals pertaining to

Leishmania focus on the promastigote form of this parasite. Given the fact that promastigotes are responsible for only the first several minutes of this chronic infection, this situation is deemed to be unfortunate. Clearly more work on this developmental form is warranted.

Many aspects of amastigote uptake into macrophages resemble classical receptor-mediated phagocytosis. Parasite uptake requires energy expenditure by macrophages but not by the parasite, and non-viable fixed amastigotes are taken up nearly as efficiently as viable organisms. In contrast, treating macrophages to prevent energy metabolism or disrupting actin polymerization prevents amastigote uptake (54). The uptake of amastigotes by macrophages involves the co-localization of cellular f-actin, paxillin, and talin to phagocytic cups that are formed around amastigotes during internalization (54). Treatment of macrophages with genestein, to inhibit protein phosphorylation, prevents the phagocytosis of amastigotes, indicating that like IgG phagocytosis this process depends on protein tyrosine phosphorylation. Thus, many features of amastigote uptake by macrophages resemble classical receptor-mediated phagocytosis.

There are several features of amastigote phagocytosis which appear to be subtly different from the uptake of IgG erythrocytes. First, amastigotes bind avidly to macrophages, and this binding is far more efficient than either complement or IgG-coated erythrocytes. Second the bound organisms are rapidly internalized by macrophages and very few organisms remain externally bound. This internalization is at least as rapid as that observed with IgG erythrocytes and more efficient, despite what appears to be a decrease in the phosphorylation of proteins during amastigote phagocytosis relative to IgG phagocytosis (54). Finally, although IgG erythrocytes have been previously reported to trigger a respiratory burst and induce cytokine production (55), amastigote phagocytosis triggers only a modest production of superoxide from macrophages (25) and little in the way of a cellular cytokine response.

Another aspect of amastigote phagocytosis that is distinctly different from that observed with promastigotes is the relative importance of exogenous host opsonins for parasite recognition by macrophages. Lesion-derived amastigotes bind efficiently to macrophages in the absence of exogenous opsonins provided by serum. In fact the addition of fresh serum does little to improve amastigote adhesion to macrophages. This is in stark contrast to the promastigote form whose adhesion to macrophages is markedly improved by serum (see above). The amastigote surface is protein-deficient relative to the high percentage of glycolipids which are of both host and parasite origin (56). To date a single dominant parasite adhesin that mediates amastigote recognition by macrophages has not been identified. Instead, there appears to be several candidate adhesins all of which may participate in amastigote recognition by macrophages. Multiple ligands may not be unexpected, given the efficiency with which this organism is recognized by macrophages. Similar to the situation with promastigotes, there is no reason to expect that all species will express the same ligands involved in macrophage recognition. Furthermore, differences in ligand expression may contribute to differences in tissue tropism exhibited by the various species.

Several years ago, we identified a heparin binding activity on amastigotes which could mediate parasite adhesion to cellular proteoglycans containing heparan sulfate (57). Parasites failed to adhere to nonmacrophage cells that lacked these proteoglycans, and parasite binding to wild-type cells could be inhibited by adding soluble heparan sulfate. This observation was somewhat unexpected because the majority of mammalian cells, including a wide variety of non-macrophage cells express proteoglycans. This observation would argue against the macrophage-specificity of amastigote adhesion, and in fact that is exactly what we observed in vitro. We showed that amastigotes adhered efficiently to a number of different nucleated cells and the degree of adhesion correlated with proteoglycan expression. However on non-macrophage cells proteoglycan adhesion was non-productive and failed to result in parasite internalization. Adhesion to macrophages, in contrast, resulted in rapid amastigote internalization, as discussed above. Thus at one level, the specificity of amastigotes for macrophages resides not in the adhesion step, but rather with phagocytosis-competence of macrophages.

Kelleher and colleagues took similar approaches to the ones described above to indicate that Leishmania lipophosphoglycan is involved in the direct adhesion of *L. major* amastigotes to macrophages (58). Soluble lipophosphoglycan could compete with whole parasites for adhesion to macrophages and a monoclonal antibody to LPG could inhibit parasite binding. In contrast to promastigote adhesion via LPG, amastigote adhesion could be inhibited by LPG from a number of different *Leishmania* species, indicating that a conserved portion of LPG may contribute to amastigote adhesion to macrophages. The similarilty between the glycan cores of LPG and the amastigote-abundant glycoinositolphospholipids (GIPLs) suggest that these later structures may also mediate amastigote adhesion, via a mechanism that is smilar to the one described for LPG. A definitive role for GIPLs in parasite adhesion has not been formally demonstrated, although there is strong evidence for a role for GIPLs in modifying the phagocytophorous vacuole of amastigotes (59) and interfering with parasite killing (60,61). The role of neutral glycosphingolipids to the adhesion of *L. amazonensis* was demonstrated by showing that antibodies to these structures could inhibit amastigote adhesion to macrophages (62). A number of other direct adhesions have been identified, but conformation of their role in amastigote infectivity remains lacking. These include a ligand for the Mannose 6 phosphate receptor (63), and mannosylated proteins which may adhere to the mannose receptor (64). The scavenger receptor may also be a candidate receptor participating in amastigote adhesion.

Several years ago Belosevic and colleagues made the observation that lesion-derived *Leishmania* amastigotes have host-derived IgG on their surface (65). This observation was confirmed by ourselves (66) and another group (67), and extended to demonstrate that IgG on the surface of amastigotes was opsonic and allowed these organisms to specifically interact with the macrophage Fcγ receptors. To determine whether host IgG is able to access intracellular amastigotes within macrophages, we performed intracellular staining for IgG on the surface of intracellular axenically grown amastigotes of *L. amazonensis*. Intracellular amastigotes residing in monolayers that were incubated in non-immune serum failed to exhibit IgG on their surface. In contrast, monolayers incubated with serum from mouse previously infected with *L. amazonensis* were strongly positive for IgG-coated intracellular amastigotes (Figure 2). These studies show that host IgG is able to access the phagolysosome and adsorb to *Leishmania* amastigotes. The antibody appears to be antigen-specific because non-immune IgG fails to opsonize intracellular amastigotes. Thus, amastigotes become opsonized with host IgG while they reside within macrophage phagolysosomes.

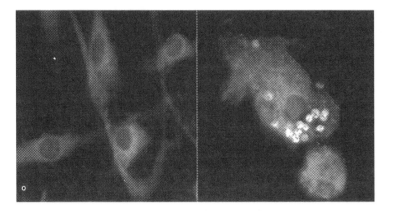

Figure 2. *Infected macrophages were incubated in 10% nonimmune serum (left) or serum from an infected mouse (right) for 2 hours. Monolayers were permeabilized and stained with a FITC-conjugated antibody to mouse IgG. Intracellular amastigotes on the right stain white in this photo.*

The biological role for surface IgG on amastigotes has been examined. Previous reports (65, 67) have suggested that host IgG was an opsonin whose role was to mediate parasite adhesion to macrophages. We were unable to demonstrate a dramatic defect in the adhesion of organisms lacking host IgG and macrophages lacking Fcγ receptors were amply able to bind amastigotes. These negative observations prompted us to look for an alternative role for surface IgG on amastigotes. We demonstrated that FcγR ligation by amastigotes exerted a profound influence on macrophage cytokine production (66). We had previously demonstrated that the ligation of phagocytic receptors on macrophages with defined particles, such as IgG-opsonized erythrocytes not only prevented the production of IL-12 (68) but it also induced the production of IL-10 (69). We proposed that this antiinflammatory cytokine mieux would have the potential to inhibit the production of a Type 1 immune response and prevent macrophage activation. We demonstrated that IgG on *Leishmania* amastigotes is a potent inducer of IL-10 from infected cells (67). We showed that this production of IL-10 was mediated by IgG on the surface of amastigotes because axenically grown amastigotes, which lacked surface IgG failed to induce the production of IL-10. Importantly, macrophages taken from mice genetically deficient in the common γ chain of the FcγR produced detectable albeit lower amounts of IL-10. We interpret this later observation to indicate that FcγR ligation by amastigotes is a potent way to induce IL-10 production, however it may not be the only stimulus for IL-10 production by lesion amastigotes.

The importance of IL-10 induction by amastigotes was addressed by both in vivo and in vitro studies. In vivo studies were performed in mice lacking IL-10. These mice were highly resistant to *L. major* infection, even on the genetically susceptible BALB/c background (67). The same applies to infections with *L. donovani* where mice lacking IL-10 exhibit a dramatic decrease in infection (70).

In vitro studies confirmed these in vivo observations. The IL-10 that was produced by infected macrophages completely abrogated the production of IL-12 and dramatically reduced TNF production by macrophages. Monolayers of macrophages treated with either recombinant IL-10 or supernatants from infected macrophages were permissive for Leishmania amastigote replication even in the presence of high quantities of IFN-γ. These studies indicate that amastigotes of *Leishmania* exploit the antiinflammatory properties of FcγR ligation and induce the production of IL-10, which renders macrophages refractory to IFN-γ activation. It should be noted, however, that mice lacking IgG are only slightly less susceptible to *L. amazonensis* infection than are wild-type mice (71). This observation, especially when contrasted with the profound resistance of IL-10-deficient mice, suggests that *Leishmania* amastigotes may have other mechanisms for inducing IL-10 production during infection.

SUMMARY

Leishmania must gain rapid entry into mammalian cells in order to establish and maintain infection. Given the importance of establishing this intracellular niche, it is therefore not surprising that both developmental forms of the parasite have adapted to utilize multiple different mechanisms to adhere to macrophages. These mechanisms include both direct adhesins and opsonin-dependent processes. The receptors mediating the direct adhesion of the two forms may be distinct, and the two forms may also take differential advantage of different host-derived opsonins. The promastigote form appears to depend on complement fixation to maximally adhere to macrophages, whereas the amastigote form exploits potential antiinflammatory effects of macrophage FcγR ligation to inhibit macrophage activation thereby facilitating its intracellular survival.

REFERENCES

1.Brittingham A, Morrison C J, McMaster W R, McGwire B. S, Chang K P, and Mosser, DM. 1995. Role of the *Leishmania* surface protease gp63 in complement fixation, cell adhesion, and resistance to complement-mediated lysis. J Immunol 155: 3102-3111.

2.Brittingham A, and Mosser DM. 1996. Exploitation of the complement system by *Leishmania* promastigotes. Parasitology Today 12: 444-447.

3.Mosser DM and Edelson PJ. 1984. Activation of the alternative complement pathway by *Leishmania* promastigotes: parasite lysis and attachment to macrophages. J Immunol 132: 1501-1505.

4.Mosser DM, Burke SK, Coutavas EE, Wedgwood JF, and Edelson PJ. 1986. *Leishmania species*: Mechanisms of complement activation by five strains of promastigotes. Exp Parasitol 2: 394-404.

5.Pearson RT and Steigbigel RT. 1980. Mechanism of lethal effect of human serum upon *Leishmania donovani*. J Immunol 125: 2195-2201.

6.Puentes SM, Sacks DL, Da Silva RP, and Joiner KA. 1988. Complement binding by two developmental stages of *Leishmania major* promastigotes varying in expression of a surface lipophosphoglycan. J Exp Med 167: 887-902.

7.Dominguez M and Torano A. 1999. Immune adherence-mediated opsono-phagocytosis: the mechanism of Leishmania infection. J Exp Med 189: 25-35

8.Pritchard DG, Volanakis JE, Slutsky GM, and Greenblatt CL. 1985. C-reactive protein binds leishmanial excreted factors. Proc Soc Exp Bio Med 178: 500-503.

9.Green PJ, Feizi T, Stoll MS, Thiel S, Prescott A, and McConville MJ. 1994. Recognition of the major cell surface glycoconjugates of *Leishmania* parasites by the human serum mannan-binding protein. Mol Biochem Parasitol 66: 319-328.

10.Claus DR, Siegel J, Petras K, Osmand AP, and Gewurz, H. 1977. Interaction of C-reactive protein with the first component of human complement. J Immunol 119: 187-196.

11.Ikeda K, Sannoh T, Kawasaki N, Kawasaki T, and Yamashina I. 1987. Serum lectin with known structure activates complement through the classical pathway. J Biol Chem 262: 7451-7454.

12.Mosser DM, Springer TA, and Diamond MS. 1992. *Leishmania* promastigotes require opsonic complement to bind to the human leukocyte integrin Mac-1 (CD11b/CD18). J Cell Biol 116: 511-520.

13.Puentes SM, Sacks DL, Da Silva RP, and Joiner, KA. 1988. Complement binding by two developmental stages of *Leishmania major* promastigotes varying in expression of a surface lipophosphoglycan. J Exp Med 167: 887-902.

14.Da Silva RP, Hall FB, Joiner KA, and Sacks DL. 1989. CR1, the C3b receptor, mediates binding of infective *Leishmania major* promastigotes to human macrophages J Immunol 143: 617-622.

15.Mosser DM and Edelson PJ. 1985. The mouse macrophage receptor for C3bi (CR3) is a major mechanism in the phagocytosis of leishmania promastigotes J Immunol 135: 2785-2789.

16.Blackwell JM, Ezekowitz RAB, Roberts MB, Channon JY, Sim RB, and Gordon S. 1985. Macrophage complement and lectin-like receptors bind *Leishmania* in the absence of serum. J Exp Med 162: 324-331.

17.Schönlau F, Scharffetter-Kochanek K, Grabbe S, Pietz B, Sorg C, and Sunderkotter C. 2000. In experimental leishmaniasis deficiency of CD18 results in parasite dissemination associated with altered macrophage functions and incomplete Th1 cell response. Eur J Immunol 30:2729-2740.

18.Rosenthal LA, Sutterwala FS, Kehrli ME, and Mosser DM. 1996. *Leishmania major*-human macrophage interactions: cooperation between Mac-1 (CD11b/CD18) and complement receptor type 1 (CD35) in promastigote adhesion. Infect Immun 64: 2206-2215.

19.Wozencraft AO, Sayers, G, and Blackwell, JM. 1986. Macrophage type 3 complement receptors mediate serum-independent binding of *Leishmania donovani*: detection of macrophage-derived complement on the parasites surface by immunoelectron microscopy. J Exp Med 164: 1332-1337.

20.Whaley K. 1980. Biosynthesis of the complement components and the regulatory components of the alternative complement pathway by human peripheral blood monocytes. J Exp Med 151: 501-516.

21.Ezekowitz AB, Sim, RB, Hill M., and Gordon, S. 1983. Local opsonization by secreted macrophage complement components. J Exp Med 159: 244-260.

22.Mosser DM and Edelson PJ. The third component of complement (C3) is responsible for the intracellular survival of *Leishmania major*. 1987. Nature 327: 329-331.

23.Sacks DL. 1992. The structure and function of the surface lipophosphoglycan on different developmental stages of *Leishmania* promastigotes, Infect Agents Dis 1: 200-206.

24.Zehavi U, El-On J, Pearlman E, Abrahams JC, and Greenblatt CL. 1983. Binding of *Leishmania* promastigotes to macrophages. Zeitschrift fur Parasitenkunde 69: 405-414

25.Channon JY, Roberts MB, and Blackwell JM. 1983. A study of the differential respiratory burst elicited by promastigotes and amastigotes of *Leishmania donovani* in murine peritoneal macrophages. Immunology 53: 345-355.

26.Wilson ME and Pearson RD. 1986. Evidence that *Leishmania donovani* utilizes a mannose receptor on human mononuclear phagocytes to establish intracellular parasitism. J Immunol 136: 4681-4688.

27.Mosser DM and Handman, E. 1992. Treatment of murine macrophages with interferon-γ inhibits their ability to bind *Leishmania* promastigotes. J Leukoc Biol; 52: 369-376.

28.Thorton BP, Vetvika V, Pitman M, Goldman RC, and Ross GD. 1996. Analysis of the sugar specificity and molecular location of the beta-glucan binding site of complement receptor type 3 (CD11b/CD18). J Immunol 156: 1235-46.

29.Mosser DM, Vlassara H, Edelson PJ, and Cerami A. 1987. *Leishmania* promastigotes are recognized by the macrophage receptor for advanced glycosylation end products. J Exp Med 165: 140-145.

30. Brownlee M, Vlassara H, and Cerami A. 1984. Nonenymatic glycosylation and the pathogenesis of diabetic complications. Ann Int Med 101: 527-537.

31. Vlassara H, Brownlee M, and Cerami A. 1985. High affinity receptor-mediated uptake and degradation of glucose-modified proteins: A potential mechanism for the removal of senescent macromolecules. Proc Natl Acad Sci USA 82: 5588-5592.

32. Handman E, Greenblatt CL, and Goding JW. 1984. An amphipathic sulfated glycoconjugate of *Leishmania*: characterization with monoclonal antibodies. EMBO J 3: 2301-2306.

33. Turco SJ, Wilkerson MA, and Clawson DR. 1984. Expression of an unusual acid glycoconjugate in *Leishmania donovani*. J Biol Chem 259: 3883-3889.

34. Handman E and Goding JW. 1985. The *Leishmania* receptor for macrophages is a lipid-containing glycoconjugate. EMBO J 4: 329-336.

35. Kelleher M, Bacic A, and Handman E. 1992. Identification of a macrophage-binding determinant on lipophosphoglycan from *Leishmania major* promastigotes. Proc Natl Acad Sci USA 89: 6-10.

36. Ilg T. 2000. Lipophosphoglycan is not required for infection of macrophages or mice by Leishmania mexicana. EMBO J 19: 1953-62.

37. Chang CS and Chang KP. 1986. Monoclonal antibody affinity purification of a *Leishmania* membrane glycoprotein and its inhibition of leishmania-macrophage binding. Proc Natl Acad Sci USA 83: 100-104.

38. Russell DG and Wilhelm, H. 1986. The involvement of the major surface glycoprotein (gp63) of *Leishmania* promastigotes in attachment to macrophages. J Immunol 136: 2613-2620.

39. Chen DQ, Kolli BK, Yadava N, Lu HG, Gilman-Sachs A, Peterson DA, and Chang KP. 2000. Episomal expression of specific sense and antisense mRNAs in Leishmania amazonensis: Modulation of gp63 level in promastigotes and their infection of macrophages in vitro. Infect Immun 68: 80-86.

40. Russell DG and Wright SD. 1988. Complement receptor type 3 (CR3) binds to an Arg-Gly-Asp-containing region on the major surface glycoprotein, gp63, of leishmania promastigotes. J Exp Med 168: 279-292.

41. Miller RA, Reed SG, and Parsons M. 1990. *Leishmania* gp63 molecule implicated in cellular adhesion lacks an Arg-Gly-Asp sequence. Mol Biochem Parasitol 39: 267-274.

42. Taniguchi-Sidle A and Isenman DE. 1992. Mutagenesis of the Arg-Gly-Asp triplet in human complement component C3 does not abolish binding of iC3b to the leukocyte integrin complement receptor type III (CR3, CD11b/CD18). J Biol Chem 267: 635-643.

43. Brittingham A, Chen G, McGwire BS, Chang KP, and Mosser DM. 1999. Interaction of Leishmania gp63 with cellular receptors for fibronectin. Infection and Immunity 67: 4477-4484.

44. Rizvi FS, Ouaissi MA, Marty B, Santoro F, and Capron A. 1988. The major surface protein of *Leishmania* promastigotes is a fibronectin-like molecule. Eur J Immunol 18: 473-476.

45. Soteriadou KP, Remoundos MS, Katsikas MC, Tzinia AK, Tsikaris V, Sakarellos C, and Tzartos SJ. 1992. The Ser-Arg-Tyr-Asp region of the major surface glycoprotein of *Leishmania* mimics the Arg-Gly-Asp-Ser cell attachment region of fibronectin. J Biol Chem 267: 13980-13985.

46. Wyler DJ, Sypek JP, and McDonald JA. 1985. In vitro parasite-monocyte interactions in human leishmaniasis: possible role of fibronectin in parasite attachment. Infect Immun 49: 305-311.

47. Pierschbacher MD and Ruoslahti E. 1984. Cell attachment activity of fibronectin can be duplicated by small synthetic fragments of the molecule. Nature 309: 30-33.

48. Wayner EA, Garcia-Pardo A, Humphries MJ, McDonald JA, and Carter G. 1989. Identification and characterization of the T lymphocyte adhesion receptor for an alternative cell attachment domain (CS-1) in plasma fibronectin. J Cell Biol 109:1321–1330.

49. Pommier CG, Inada S, Fries LF, Takahashi T, Frang MM, and Brown EJ. 1983. Plasma fibronectin enhances phagocytosis of opsonized particles by human peripheral blood monocytes. J Exp Med 157:1844–1854.

50. Bohnsack JF, O'Shea JJ, Takahashi T, and Brown EJ. 1985. Fibronectin enhanced phagocytosis of an alternative pathway activator by human culture-derived macrophages is mediated by the C4b/C3b complement receptor (CR1). J Immunol 135:2680–2686.

51. Savoia D, Biglino S. Cestaro A, and Zucca M. 1992. Effect of fibronectin and interferon-gamma on the uptake of *Leishmania major* and *Leishmania infantum* promastigotes by U937 cells. Microbiologica 15:51–56.

52. Hodgkinson VH, Soong L, Duboise SM, and McMahon-Pratt D. 1996. Leishmania amazonensis: cultivation and characterization of axenic amastigote-like organisms. Exp Parasitol 83:94-105.

53. Handman E. 1983. Association of serum proteins with cultured Leishmania: a warning note. Parasite Immunol 5:109-112.

54. Love DC, Kane MM and Mosser DM. 1998. Leishmania amazonensis: the phagocytosis of amastigotes by macrophages. Exp. Parasitol 88:161-171.

55. Swanson JA and Baier SC. 1995. Phagocytosis by zippers and triggers. Trends in Cell Biol 5:89-92.

56. McConville MJ and Ferguson MA. 1993. The structure, biosynthesis and function of glycosylated phosphatidylinositols in the parasitic protozoa and higher eukaryotes. Biochem J 294:305-324.

57. Love DC, Esko JD, and Mosser DM. 1993. A heparin-binding activity on Leishmania amastigotes which mediates adhesion to cellular proteoglycans. J Cell Biol 123: 759-766.

58. Kelleher M, Moody S F, Mirabile P, Osborn AH, Bacic A, and Handman E. 1995. Lipophosphoglycan blocks attachment of *Leishmania major* amastigotes to macrophages. Infect. Immun. 63: 43-50.

59. Peters C, Stierhof Y, and Ilg T. 1997. Proteophosphoglycan secreted by Leishmania mexicana amastigotes causes vacuole formation in macrophages. Infect Immun 65:783-786.

60. Winter G, Fuchs M, McConville MJ, Stierhof YD, and Overath P. 1994. Surface antigens of Leishmania mexicana amastigotes characterization of glycoinositol phospholipids and a macrophage-derived glycosphingolipid. J Cell Sci 107:2471-2482

61. Proudfoot L, O'Donnell CA, and Liew FY. 1995. Glycoinositolphospholipids of Leishmania major inhibit nitric oxide synthesis and reduce leishmanicidal activity in murine macrophages. Eur J Immunol 25:745-750.

62. Straus AH, Levery SB, Jasiulionis MG, Salyan ME, Steele SJ, Travassos LR, Hakomori S, and Takahashi HK. 1993. Stage-specific glycosphingolipids from amastigote forms of *Leishmania* (L.) *amazonensis*. Immunogenicity and role in parasite binding and invasion of macrophages. J Biol Chem 268: 13723-13730.

63. Saraiva EM, Andrade AF, and de Souza W. 1987. Involvement of the macrophage mannose-6-phosphate receptor in the recognition of Leishmania mexicana amazonensis. Parasitol Res 73:411-416.

64. Chakraborty P and Das PK. 1988. Role of mannose/N-acetylglucosamine receptors in blood clearance and cellular attachment of leishmania donovani. Mol Biochem Parasitol 28:55-62.

65. Guy R A. and Belosevic M. 1993. Comparison of receptors required for entry of *Leishmania major* amastigotes into macrophages. Infect. Immun. 61: 1553-1558.

66. Mentink Kane, M. and Mosser, D.M. 2001. The role of IL-10 in promoting disease progression in Leishmaniasis. J. Immunology 166:1141-47.

67. Peters C, Aebischer T, Stierhof YD, Fuchs M, and Overath P. 1995. The role of macrophage receptors in adhesion and uptake of Leishmania mexicana amastigotes. J Cell Sci 108:3715-3724.

68. Sutterwala FS, Noel GJ, Clynes R, and Mosser DM. 1997. Selective suppression of interleukin 12 induction following macrophage receptor ligation. J. Exp Med 185:1977-1985.

69. Sutterwala FS, Noel GJ, Salgame PS, and Mosser DM. 1998. Reversal of proinflammatory responses by ligating the macrophage FcγRI. J Exp Med 188:217-222.

70. Murphy ML, Wille V, Villeges EN, Hunter CA, and Farrell JP. 2001. IL-10 mediates susceptibility to L. donovani infection. Eur J Immunol 31:2848-2856.

71. Kima PE, Constant SL, Hannum L, Colmenares M, Lee KS, Haberman AM, Shlomchik MJ, and McMahon-Pratt D. 2000. Internalization of Leishmania mexicana complex amastigotes via the Fc receptor is required to sustain infection in murine cutaneous leishmaniasis. J Exp Med 191:1063-1068.

LEISHMANIA INFECTION AND MACROPHAGE FUNCTION

Greg Matlashewski

Department of Microbiology and Immunology, McGill University, Montreal Canada

INTRODUCTION

Leishmania have a digenetic life cycle, alternating between free living flagellated promastigotes in sand flies to obligate intracellular aflagellated amastigotes in macrophages. It is important and relevant to define the molecular basis for this host parasite interaction because this is a chronic infection affecting over two million people each year (reviewed in 1,2). *Leishmania* infections are responsible for a wide range of human pathologies ranging from scar yielding cutaneous lesions to fatal visceral infections (1,2). In order to appreciate the molecular basis for this infection, it is necessary to first consider the macrophages since these are the only cells in which *Leishmania* amastigotes proliferate in the mammalian host. An obvious question is how *Leishmania* amastigotes survive in the hostile environment of macrophages since these cells have among the most potent antimicrobial properties of any cell type in the body. As with many intracellular infectious agents including viruses and bacterial pathogens, *Leishmania* must transform the cell they infect. Host macrophages become governed by the *Leishmania* intruder, instead of obeying the normal rules necessary for macrophages to fulfill their assigned role in defense against infections. In other words, *Leishmania* causes the macrophage to alter allegiances if it is to survive, proliferate, and ultimately be transmitted to a new host. This chapter will review some of the major advances that have been made in defining the molecular mechanisms which allow this pathogen to survive and proliferate in macrophages. This information is valuable for the developing new strategies to treat leishmaniasis that could also have important implications for the treatment of other serious infections which target macrophages as their host cell.

INFECTION OF THE HOST MACROPHAGES

Macrophages are present in all tissues in the body and carry out fundamental protective functions against invading pathogens through phagocytosis and destruction of microorganisms and this is part of the innate immunity (reviewed in 3,4). Macrophages also play an essential role in acquired immunity to infectious organisms through presenting antigen to lymphocytes and secreting a variety of cytokines during the development of the immune response. With respect to *Leishmania*, the infection process begins with the bite of the infected sandfly resulting in the inoculation of promastigotes into the dermis. The promastigotes adhere to resident or recruited macrophages through a variety of receptors, the major ones being complement receptors type I and type III (5). Complement-dependent promastigote adhesion to the macrophage is followed by internalization by phagocytosis (6). The promastigote containing phagosome subsequently undergoes a maturation process involving the fusion with endocytic organelles including endosomes and lysosomes resulting in the mature phagolysosome (7). The mature phagolysosome is the major microbicidal site in macrophages due to the presence of lysosomal hydrolysing enzymes, acidic pH, and toxic molecules such as nitric oxide (NO) and other reactive oxygen intermediates. However, *Leishmania donovani* promastigotes appear to slow the process of phagolysosome maturation through impairing the fusion of the parasite containing phagosome with the endosomes and lysosomes (8). The inhibition of phagolysosomal maturation likely represents a survival strategy used by the invading promastigotes since they are sensitive to low pH. Delaying the maturation process of the phagolysosome allows sufficient time for the differentiation from promastigotes to amastigotes to occur which takes about 48 hours following phagocytosis. The amastigotes are able to survive in the acidic pH of the mature phagolysosome and indeed require the low pH to begin proliferation and to express amastigote specific genes (9). Following amastigote establishment and multiplication, the infected macrophage eventually will lyse releasing new amastigotes which are rapidly taken up by surrounding macrophages. Since amastigotes require the acidic environment for multiplication, they do not delay the maturation of the parasitized phagolysosome once they have been taken up by macrophages. In this regard, it is noteworthy that the promastigote surface molecule lipophosphoglycan (LPG) appears to be responsible for delaying phagolysosome maturation (8).

Amastigotes represent the major part of the life cycle since they are present within the mammalian host macrophage phagolysosomes for anywhere from months to many years. By comparison, promastigotes are only present in the sandfly host for a few days and serve only to transfer the parasite to a new mammalian host. The successful transfer from the sandfly host to the human host is a perilous and difficult journey and it is likely that only a few of the 100-200 sand fly-inoculated promastigotes survive the journey and safely make it to viable amastigotes.

Survival in Macrophages through Impairment of the Acquired Immune Response.

Following the initial infection process and survival of the innate immune response, the amastigotes must next survive the acquired immune response in order to establish a long-term chronic infection. Macrophages play a central role in initiating the acquired immune response to infectious pathogens through presenting antigen and releasing regulatory cytokines to T cells. In order to resolve the *Leishmania* infection, the acquired immune response must include the development of a T helper type 1 (Th1) immune response such that the T-helper cells are capable of producing IFN-γ which subsequently "Activates" macrophage microbicidal activity (10). IFN-γ is the most potent macrophage activating cytokine and is essential to induce macrophage killing of amastigotes (11, 12). Macrophage activation refers to the conversion of the normal resting macrophage to the activated state where macrophages acquire a number of functional changes including the release of microbicidal compounds such as nitric oxide (NO), reactive oxygen intermediates, and activation of hydrolytic lysosomal enzymes (13). Reactive oxygen intermediates (14) and NO (15) are particularly toxic to *Leishmania* and thus activated macrophages are capable of killing amastigotes and resolving the infection.

Leishmania has developed several mechanisms to suppress the development of the Th1 immune response and the production of IFN-γ to circumvent macrophage activation. Infected cells are defective in the expression of MHC class II molecules on the surface (16) and co-stimulatory molecules (17). In this manner, infected macrophages have a reduce capacity to directly interact with T cells thus impairing the initiation of the acquired immune response. Another important defence mechanism is the inhibition of IL-12 production by *Leishmania* infected macrophages (18). The release of IL-12 by the infected macrophage is essential for initiating the Th1 response. Therefore the inhibition or circumvention of IL-12 induction during the establishment of infection is an important adaptive strategy to delay the onset of the acquired immune response. The mechanism in which *Leishmania* inhibits the release of IL-12 is unknown however impairment at both the transcriptional and post-transcriptional levels of the p40 subunit gene of IL-12 have been described (19, 20). Studies have also shown that *Leishmania* infected macrophages have several signal transduction pathways which are inhibited including the protein kinase C (PKC) (21,22,23) and JAK/STAT (24, 25) signaling networks and this could be either directly or indirectly involved with the impairment of IL-12 release by the infected cells. Taken together, the overall impairment of the acquired immune response through uncoupling the communication between the infected macrophage and T-cells is an important strategy to circumvent macrophage activation.

Impairment of IFN-γ Mediated Signal Transduction during Macrophage Infection.

As discussed above, IFN-γ is the most potent cytokine for the induction of macrophage activation for leishmaniacidal and microbicidal activity. Following the binding of IFN-γ to it receptor, the receptor dimerizes and stimulates the phosphorylation of two receptor associated tyrosine kinases, JAK1 and JAK2 which subsequently phosphorylate other cytoplasmic transcription factors including STAT1. Phosphorylated STAT1 then dimerizes and translocates into the nucleus where it stimulates the transcription of IFN-γ responsive genes, some of which express proteins required for macrophage activation. Since this pathway is important for macrophage activation, *Leishmania* have further developed mechanisms to suppress the ability of macrophages to respond to IFN-γ For example, it has been revealed that infected macrophages have a reduction in phosphorylation of JAK1, JAK2, and STAT1 following IFN-γ stimulation (24, 25, 26). Recent evidence suggests that the reduced phosphorylation of these signal transduction intermediates in infected cells could be due to the activation of phosphotyrosine phosphatases which removes phosphate groups from tyrosine residues (27, 28). In this manner, the IFN-γ signaling pathway is compromised impairing the activation of macrophages and ensuring the survival of the *Leishmania* amastigotes.

Enhancement of Macrophage Viability during Infection

As outlined above, *Leishmania* infection suppresses a number of specific macrophage activities associated with signal transduction and activation. However, it is necessary to appreciate that infection also enhances various macrophage activities which further secures the appropriate host environment. For example, it was revealed that primary infected macrophages could survive longer in the absence of the macrophage growth factor, colony stimulating factor 1 (CSF-1) than could uninfected macrophages (29). Furthermore, it was demonstrated that infection was able to inhibit macrophage apoptosis caused by the removal of CSF-1 (30). Infected cells express several genes which could maintain macrophage survival in the absence of exogenous growth factor including granulocyte-monocyte colony stimulating factor (GM-CSF) (30). Taken together, these observations revealed that infection can have a short term positive effect on macrophage survival through inhibiting apoptosis. Clearly, it is advantageous to keep the infected cell viable long enough to ensure that the amastigotes can replicate in its host cell and this is a strategy which is also commonly observed in virally infected cells (31). Infection with *L. donovani* also induces the expression the macrophage inflammatory protein genes 1α and 1β (MIP-1α and MIP-1β) in macrophages (32). This is an interesting observation because one of the major functions of the MIPs is to mediate the recruitment of monocytes/macrophages to the site of infection. This therefore represents another example of how infection can enhance certain macrophage functions to potentially augment the spread of infection through recruiting more host macrophages. It is interesting to note that a mouse model has been described in which the MIP-1_ genes have been experimentally removed through gene knockout technology, and coincidentally, *Leishmania* infection was used to characterize the immunological phenotype of these mice (33). As might be predicted, the MIP-1α knockout mice had lower levels of infection than the control mice. This is consistent with the observation that *L. donovani* infection of macrophages results in the induction of MIP-1α and MIP-1β gene expression, and in the absence of these genes, the spread of infection is limited.

Macrophage Gene Expression during *Leishmania* Infection

The above described studies have been very informative through focusing on a limited number of specific macrophage biochemical pathways during infection. However, another relevant question is what effect does infection have on global macrophage gene expression. It is possible to address this issue through microarray analysis to compare mRNA levels for hundreds of genes in infected and non-infected cells to identify macrophage genes which are induced or suppressed during infection and which may therefore influence the outcome of infection. In the first study to undertake this approach, it was revealed that *L. donovani* infection resulted in an overall suppression of macrophage gene expression (32). About 40 percent of the expressed genes in macrophages have a reduced expression during *L. donovani* infection. Only a few genes had their expression increased during infection including the MIP genes described above (32). Such suppression of gene expression likely ensures that infected macrophages remain in a quiescent state. It will be interesting in future similar studies to determine how different species of *Leishmania* which cause fatal visceral or non-fatal cutaneous infections effect the overall expression of macrophage genes. It will be important to define the mechanism in which *Leishmania* infection inhibits expression of some but not all macrophage genes.

Treatment of Infection through Manipulation of Macrophage Function

The most important objective of defining the molecular basis for macrophage infection with *Leishmania* is to use this information to develop effective strategies for treating this infection. Based on the above information, it is clear that infected macrophages behave differently than non-infected macrophages but is it possible to activate infected macrophages to become leishmaniacidal? This has been attempted with recombinant IFN-γ. Treatment with recombinant IFN-γ was first attempted in combination with antimony therapy for visceral leishmaniasis and it was observed that seven out of nine cases of antimony-resistant patients could be cured with this combination (34). It was also revealed that the combination of IFN-γ and antimony was able to shorten the time needed to eliminate the infection in previously untreated patients (35). A subsequent trial however revealed that IFN-γ treatment alone for visceral leishmaniasis was only partially effective (36) and was not effective in the treatment of cutaneous leishmaniasis (37). Based on these and other studies, IFN-γ does not appear to be sufficiently active to be used alone to treat leishmaniasis, but does augment the efficacy of antimony therapy. Because of limited success, high cost, and side effects associated with IFN-γ, it is not likely to be widely used for the treatment of leishmaniasis.

There have also been some advances made in the development of small molecules which are capable of activating macrophages. For example, imidazoquinoline compounds (imiquimod and S28463) have been shown to stimulate monocytes to release a number of cytokines associated with macrophage activation (38, 39). It has also been demonstrated that imidazoquinolines are capable of inducing the expression of a key leishmaniacidal gene encoding the inducible nitric oxide synthase enzyme (iNOS) which synthesises nitric oxide (NO) from L-arginine (39, 40). Imidazoquinoline was further shown to induce the synthesis of NO in *Leishmania* infected macrophages *in vitro* and this was associated with potent leishmaniacidal activity (40). Moreover, topical application of imiquimod was capable of significantly impairing the development of cutaneous leishmaniasis in BALB/c mice (40). Additional studies using microarray analysis revealed that imidazoquinolines selectively induced the expression of a relatively small number of macrophage genes involved in activation and inflammation including the iNOS, CD40, and IL-1 genes (39). An important consideration is that imidazoquinoline has been approved by the American Food and Drug Administration (FDA) for the treatment of cutaneous genital lesions caused by human papillomavirus infection (41) and therefore was available to test topically against human cutaneous leishmaniasis. Imiquimod used in combination with standard antimony treatment was effective in two respects including: (i), reducing the time it took to clear the infection and (ii), healing infections which were resistant to standard antimony therapy alone (42). As with IFN-γ, imidazoquinoline therefore appears to be most effective when used in combination with antimony. Nevertheless, because of the currently available topical formulation, absence of side effects, and lower cost, the combination therapy of imidazoquinoline and antimony may be useful for the future management of cutaneous leishmaniasis, especially in those patients who do not respond to antimony alone. Moreover, based on the *in vitro* and *in vivo* observations with imidazoquinolines in the treatment of leishmaniasis, it has also been shown to be effective at activating macrophages to kill intracellular infection with *mycobacteria* and this may have considerations for the treatment of tuberculosis (43). These types of recent advances clearly justify the many basic studies to define the molecular basis for *Leishmania* infection of macrophages.

ACKNOWLEDGEMENTS

The author acknowledges the support of the Canadian Institutes for Health Research (CIHR) and the United Nations Development Program/World Bank/World Health Organization special program for research and training in tropical diseases.

REFERENCES

1. Herwaldt, B. Leishmaniasis. Lancet 354 1999. 1191-1199.
2. Berman, J. Human Leishmaniasis: Clinical, Diagnostic, and Chemotherapeutic Developments in the Last 10 years. Clin Infect Dis 24 1997. 684-703.
3. Johnston R. 1988. Monocytes and macrophages. New Engl J Med 318: 747-752.
4. Alexander J, Russel D. 1992. The interaction of *Leishmania* species with macrophages. Adv Parasitol 31: 175-254.
5. Mosser D, Rosenthal LA 1993. *Leishmania*-macrophage interactions: multiple receptors, multiple ligands and diverse cellular responses. Sem Cell Biol 4: 315-322.
6. Brittingham A, Mosser D. 1996. Exploitation of the complement systems by *Leishmania* promastigotes. Parasitol Today 12: 444-447.
7. Antoine J-C, Prina E, Lang T, Courret N 1998. The biogenesis and properties of the parasitophous vacuoles that harbour *Leishmania* in murine macrophages. Trends Microbiol 7: 392-401.
8. Desjardin M, Descoteaux A, 1997. Inhibition of phagolysosomal biogenesis by *Leishmania* lipophosphoglycan. J Exp Med 185: 2061-2068.
9. Charest H, Zhang W, and Matlashewski G 1996. The developmental expression of *Leishmania donovani* A2 amastigote specific genes is post-transcriptionally mediated and involves elements located in the 3'UTR J Biol Chem 271: 17081-17090.
10. Reed S, Scott P 1993. T-cell and cytokine responses in leishmaniasis. Curr Biol 5: 524-531
11. Gallin J, Farber S, Holland SM, Nutman TB 1995. Interferon gamma in the management of infectious diseases. Ann Intern Med. 123: 216-223.
12. Belosevic M, Finbloom D, Van der Meide P, Slayter M, Nacy C 1989. Adminstration of monoclonal anti-IFN-gamma antibodies in vivo abrogates natural resistance of C3H/HeN mice to infection with *Leishmania major*. J Immunol 143: 266-274.
13. Bogdan C, Rollingoff M 1999. How do protozoan parasites survive inside macrophages. Parasitol Today 15: 22-28.
14. Hughes H 1988. Oxidative killing of intracellular parasites mediated by macrophages. Parasitol Today 4: 340-347.
15. Mossalayi MD, Arock M, Mazier D, Vincendeau P, Vouldoukis I 1999. The human immune response during cutaneous leishmaniasis: NO problem. Parasitol Today 15: 342-344.

112 Matlashewski

16. Reiner NE, Ng W, McMaster W 1987. Parasite-accessory cell interactions in murine leishmaniasis. *Leishmania donovani* suppresses macrophage expression of class I and class II major histocompatibility complex gene products. J Immunol 138: 1926-1932.
17. Kaye PM, Rogers NJ, Curry AJ, Scott JC 1994. Deficient expression of co-stimulatory molecules on *Leishmania* infected macrophages. Eur J Immunol 24: 2850-2854.
18. McDowell MA, Sacks D 1999. Inhibition of host cell signal transduction by *Leishmania:* observations relevant to the selective impairment of IL-12 responses. Curr Opin Microbiol 2: 438-443.
19. Piedrafita D, Proudfoot L, Nikolaev AV, Xu D, Sands W, Feng GJ, Thomas E, Brewer J, Ferguson MA, Alexander J, Liew FY 1999. Regulation of macrophage IL-12 synthesis by leishmania phosphoglycans. Eur J Immunol 29: 235-244.
20. Weinheber N, Wolfram M, Harbecke D, Aebischer T 1998. Phagocytosis of *leishmania mexicana* amastigotes by macrophages leads to a sustained suppression of IL-12 production. Eur J Immunol 28: 2467-2477.
21. Descoteaux A, Matlashewski G, Turco S, Sacks D, Matlashewski G 1992. Inhibition of macrophage protein kinase C-mediated protein phosphorylation by *Leishmania donovani* lipophosphoglycan. J Immunol 149: 3008-3015
22. Olivier M, Brownsey RW, Reiner NE 1992. Defective stimulus-response coupling in human monocytes infected with *Leishmania donovani* is associated with altered activation and translocation of protein kinase C. Pro Natl Acad Sci USA 89: 7481-7485
23. Moore K, Labrecque S, Matlashewski G. 1993. Alteraction of *L. donovani* infection levels by selective impairment of macrophage signal transduction. J Immunol 150:4457-4465.
24. Blanchette J, Racette N, Faure R, Siminovitch K, Olivier M. 1999. *Leishmania*-induced increases in activation of macrophage SHP-1 tyrosine phosphatase are associated with impaired IFN-gamma-triggered JAK2 activation. Eur J Immunol 29:3737-3744.
25. Nandan D, Reiner NE 1995. Attenuation of gamma interferon-induced tyrosine phosphorylation in mononuclear phagocytes infected with *Leishmania donovani:* selective inhibition of signaling through Janus Kinases and Stat1. Inf Immun 63:4495-4500.
26. Ray M, Gam A, Boykins R, Kenney R 2000. Inhibition of Interferon-gamma signaling by *Leishmania donovani.* J Inf Dis 181:1121-1128.
27. Olivier M, Romero-Gallo BJ, Matte C, Blanchette J, Posner BI, Tremblay MJ, Faure R 1998. Modulation of interferon-gamma-induced macrophage activation by phosphotyrosine phosphatases inhibition. Effect on murine Leishmaniasis progression. J Biol Chem 273:13944-13949.
28. Nandan D, Knutson K, Lo R, Reiner N 2000. Exploitation of host cell signalling machinery: activation of macrophage phosphotyrosine phosphatases as a novel mechanism of molecular microbial pathogenesis. J Leuk Biol 67:464-470.
29. Moore K, Turco S, Matlashewski G. 1994. *Leishmania donovani* infection enhances macrophage viability in the absence of exogenous growth factor. J Leuk Biol 55:91-98.
30. Moore K, Matlashewski G 1994. Intracellular infection by *Leishmania donovani* inhibits macrophage apoptosis. J Immunol 152:2930-2937.
31. Roulston A, Marcellus R, Branton P 1999. Viruses and apoptosis. Ann Rev Microbiol 53:577-628.
32. Buates S, Matlashewski G 2001. General suppression of macrophage gene expression during *Leishmania donovani* infection. J Immunol. 166:3416-3422.
33. Sato N, Kuziel W, Melby P, Reddick R, Kostecki V, Zhao W, Maedam N, Ahuja S 1999. Defects in the generation of IFN-gamma are overcome to control infection with *Leishmania donovani* in CC chemokine receptor (CCR 5), macrophage inflammatory protein-alpha, of CCR2-deficient mice. J Immunol 163: 5519-5525.
34. Badaro R, Falcoff E, Badaro FS 1990. Treatment of visceral leishmaniasis with

pentavalent antimony and interferon gamma. N Engl J Med 322:16-21.

35. Squires KE, Rosenkaimer F, Sherwood JA, Forni AL, Were JB, Murray HW 1993. Immunochemotherapy for visceral leishmaniasis: a controlled pilot trial of antimony versus antimony plus interferon gamma. Am J Trop Med Hyg 48:666-669.

36. Sundar S, Murray HW 1995. Effect of treatment with interferon gamma alone in visceral leishmaniasis. J Inf Dis 172:1627-1629.

37. Harms G, Chehade AK, Douba M 1991 A randomized trial comparing a pentavalent antimonial drug and recombinant interferon-gamma in the local treatment of cutaneous leishmaniasis. Trans R Soc Trop Med Hyg 85:214-216

38. Miller RL, Gerster ML, Owens HB, Slade HB, Tomai MA 1999. Imiquimod applied topically: a novel immune response modifier and a new class of drug. Int J Immunopharm 21:1-14.

39. Buates S, Matlashewski G 2001. Identification of genes induced by a macrophage activator, S-28463, using gene expression array analysis. Antimicro Ag Chemother 45:1137-1142.

40. Buates S, Matlashewski G 1999. Treatment of experimental leishmaniasis with the immunomodulators imiquimod and S-28463; efficacy and mode of action. J Inf Dis 179: 1485-1494.

41. Beutner KR, Geisse JK 1997. Imiquimod: an immune response modifier for the treatment of genital warts. Today Therap Trends 15:165-178.

42. Arevalo I, Ward B, Miller R, Ming T-C., Najar E., Alvarez E., Matlashewski G, and Llanos-Cuentas A. Successful treatment of drug resistant human cutaneous leishmaniasis using an immunomodulator, imiquimod. Clin Infect Dis in press

44. Moisan J, Wojciechowski W, Guilbault C, Lachance C, Marco S, Skamene E, Matlashewski G, and Radzioch D. Clearance of infection with *M. bovis BCG* in mice is enhanced by the treatment with S28463 and its efficiency depends on the expression of the wild type resistant allele of the *Nramp1* gene. Antimicro Ag Chemother in press

CLINICAL AND LABORATORY ASPECTS OF *LEISHMANIA* CHEMOTHERAPY IN THE ERA OF DRUG RESISTANCE

Dan Zilberstein[1] and Moshe Ephros[2]

[1]Department of Biology, [2]Department of Pediatrics, Carmel Medical Center and the [2]Faculty of Medicine, Technion-Israel Institute of Technology, Haifa, 32000, Israel

INTRODUCTION

This chapter aims to merge clinical and basic research aspects of anti-leishmanial chemotherapy. Initially, current medical practice will be described and clinical problems and challenges, including patient, geographic, parasite and drug issues, will be addressed. In the second part, knowledge about the mode of action and mechanism(s) of resistance of currently used anti-leishmanial drugs will be described.

Recently, anti-leishmanial chemotherapy for visceral (VL), cutaneous (CL), and to a certain extent, mucocutaneous (MCL) leishmaniasis has been extensively reviewed (1-6). These are timely publications, since therapeutic options have increased over the last decade or so. Systematic and high quality clinical studies have identified a group of new treatment options, particularly for VL (2,7-22). Despite major progress, treating the enormous clinical and biological spectrum of the leishmaniases has remained highly problematic. In parallel, despite investigations which have focused on the anti-leishmanial activity of pentavalent antimony, amphotericin B, pentamidine, miltefosine and others (5,23-25), and in addition, those which have focused on mechanisms of parasite drug resistance, such as multidrug resistance (MDR), gene amplification etc. (26-30), much remains unknown. The current status of these issues and future research directions will be discussed.

THE TREATMENT OF HUMAN LEISHMANIASIS
Patient issues

Leishmania parasites infect people of all ages though each species may display an age or gender predilection, e.g. VL in the middle east and Latin America affects primarily children (*L. infantum* and *L. chagasi*, respectively), while VL in India affects children and young adults (*L. donovani*); CL/MCL in parts of the new world occurs primarily in males who tend to be exposed at the worksite. Even in a specific focus inhabited by a single parasite species, the clinical spectrum of disease will vary. Anti-leishmanial drugs, however, should ideally be appropriate for all ages and also for pregnant women.

The immunological basis for the pathogenesis of leishmaniasis is largely related to whether a host will mount a Th1 or Th2 response (31,32,32-47). The former response may result in no infection or in asymptomatic to mild to moderate clinical disease, usually with good healing, with or without specific therapy. The latter T cell response is associated with severe disease or with no or partial response to therapy. Beyond the inherent genetic predilection of an individual to mount either of these responses, secondary deficiencies of (cellular) immunity affect *Leishmania* infectivity and virulence. Given that *Leishmania* endemicity occurs very often in poor, developing countries, therapy must be effective even in those patients immunocompromised by malnutrition and by concomitant chronic diseases and infections.

Importation to the U.S.A. of visceralizing *L. tropica* by veterans of the Persian Gulf War (48,49) or simply (individual) travel-related importation from endemic to non-endemic countries has challenged the concept that leishmaniasis is a disease of the developing world. Even along "affluent" Western Europe's *Leishmania*-endemic Mediterranean coast, HIV/ *Leishmania* co-infection has emerged as a major problem. Co-infected patients suffer from a more virulent HIV course, severe and atypical leishmaniasis, poorer responses to therapy, and relapses (50-56). Over the coming years HIV/*Leishmania* co-infection will probably also become a major health issue in less affluent countries e.g. India, Brazil, and parts of Africa (3).

Geographic issues

The major burden of leishmaniasis is in poor, developing countries with inadequate financial resources to deal with the (control, diagnosis and) treatment of leishmaniasis. In many if not most endemic areas, foci of leishmaniasis are rural, making access difficult. Animal reservoirs and sandfly vectors vary with geography as well. Importantly, the changing epidemiology of leishmaniasis has favored spreading endemicity: urbanization in Brazil; war-related mass population movements in Iran; post kala azar dermal leishmaniasis (PKDL) and epidemic VL in the Sudan; antimony resistance and HIV in India; HIV/AIDS and intravenous drug abuse juxtaposed with *L. donovani* exposure in southern Europe are some of the more pronounced examples.

Parasite issues

The intracellular (within the reticulo-endothelial system) location of the amastigote, the human form of the parasite, makes its study as well as the study of anti-leishmanial drugs difficult. Approximately 20 *Leishmania* species pathogenic to humans exist and leishmanial diversity extends much beyond simple old vs. new world differences (6). Parasite diversity combined with human and geographic diversity, are responsible for the highly variable clinical spectrum of leishmaniasis (VL, CL, MCL, PKDL, DCL, recidivans, etc) and for the variable responses of theses parasites to some anti-leishmanial drugs (primary drug resistance).

Drug issues

Most highly effective agents (e.g. pentavalent antimonials, amphotericin B, pentamidine) are either administered parenterally for a prolonged period of time, expensive, of limited availability, toxic, or combinations of these properties. Until recently, when miltefosine was shown to be highly effective for treating VL (10,11,18-22), oral agents (e.g. azoles, allopurinol) were found to be less effective than parenteral therapy (57-61). Combining antimony and aminosidine for antimony-resistant VL (62,63) has emphasized the potential value of concomitant treatment with multiple drugs, possibly including immunomodulatory compounds e.g. IFN-γ and GM-CSF (64-69). Neither has been studied systematically enough.

Given the relatively benign and self-resolving nature of much new and old-world cutaneous leishmaniasis (e.g. *L. major, L. tropica, L. mexicana*), and taking into account drug toxicity and expense issues, systemic and possibly even local treatment may not always be warranted. Few locally applied agents proven highly effective exist and most have not been adequately studied (4,5).

Although new drugs are on the horizon, the time to final approval for human use is not in step with current needs and certainly not with future needs, especially considering the impact of emerging resistance on current front-line agents, particularly antimony, the impact of HIV/*Leishmania* co-infection, and the spread of leishmaniasis (e.g. refugee movements, urbanization) to new and heavily populated areas with endemic potential (1-3,6).

Principles of treatment

- Base treatment decisions on **accurate laboratory diagnosis** of the infecting *Leishmania* species and not only the clinical syndrome.
- **Individualize therapy** for the patient in the context of public health (sporadic vs. epidemic, what drugs available) and cost issues (what budget is available for drugs and for hospitalization) and the patient's health status (HIV infected, other immune compromise).
- Take the **natural history of untreated infection** into account in order to decide what treatment is required: symptomatic VL is almost always lethal if untreated while *L. major* in a cosmetically and functionally unimportant site is probably a nuisance at worst.

- Choose the **most effective and least toxic agent** (and route of therapy) relevant for the infecting *Leishmania* species and clinical syndrome. Do not curtail therapy or give sub-optimal doses for fear of toxicity.
- Consider the **epidemiology of parasite drug-resistance**.
- Plan **alternative or additional therapies** in case of treatment failure (persistence or relapse) or drug toxicity. Plan chronic suppressive therapies for the immune compromised.
- Perform **appropriate follow-up**, clinical and/or laboratory, for the correct period of time.

Visceral Leishmaniasis

Until the last decade or so, therapeutic options for VL were few and a pentavalent antimony compound was almost the only option except for the more toxic pentamidine. Currently, numerous therapeutic options (new and rediscovered) are available, however, the mainstay of anti-leishmanial therapy in most endemic countries where antimony susceptibility remains high is still pentavalent antimony, this, despite cost, logistics (parenteral, long duration) and (moderate) toxicity (1-3,5,6,70-72). Antimony resistance has emerged focally worldwide, though most extensively in certain regions of Bihar, India (5,73-78). In more affluent countries, lipid-associated formulations of amphotericin B have been adopted for effective and relatively non-toxic short course therapy. Less fortunate patients will need to receive amphotericin B deoxycholate (ABD), pentamidine, both toxic, or aminosidine. ABD treatment, though relatively toxic, prolonged, and demanding in terms of laboratory follow-up, is the usual alternative in poor, endemic countries. Lipid-associated amphotericin preparations (Amphotec,Abelcet, AmBisome, Amphotericin B-fat emulsion) have become first line therapy because they can be used in short-course, sometimes single dose and usually less toxic regimens (2,3). The great expense of these drugs, to a certain extent, is offset by the shortened duration of therapy (2,3). Generic preparations of liposomal amphotericin B may provide cheap, yet effective (therefore accessible, good) alternatives in endemic countries with limited resources (79-81).

Aminosidine is a reasonably effective parenteral anti-leishmanial aminoglycoside (82,83). Lengthy protocols and potential oto- and nephrotoxicity are drawbacks. Combining aminosidine with other anti-leishmanial drugs may overcome parasite antimony resistance while shortening treatment regimens (62,63). Oral azoles, with or without allopurinol, have not proven efficacious enough for treating VL (57,59). WR6026, an orally administered 8-aminoquinoline, is under renewed evaluation (61,84-86). Recently, oral miltefosine (the membrane-active phospholipid derivative, hexadecylphosphocholine) has proven efficacious and relatively non-toxic for treating Indian VL, especially in Bihar, where antimony resistance is the rule, pentamidine failures are common, and where amphotericin B in its deoxycholate or lipid associated forms were the mainstays of therapy until the introduction of miltefosine (3,5). Effective in both adults and children and relatively non-toxic, miltefosine (originally intended to be an antineoplastic agent) cannot be used for pregnant women due to potential teratogenicity (3) . Treating VL during pregnancy remains a significant challenge (87).

The immune modulators, IFN-γ and GM-CSF, given in combination with anti-parasitic drugs (usually antimony), are expensive and not readily available, but can be efficacious (2,3). Other immunomodulatory compounds may become ancillary anti-leishmanial agents and will either stimulate the Th1 response, or down-regulate the Th2 response (3). Particularly challenging is the treatment of HIV/*Leishmania* co-infection. Poor initial responses and frequent relapses lead to the necessity of continual suppressive therapy. Highly active anti-retroviral therapy (HAART) with the ensuing immune-reconstitution may somewhat alleviate this most difficult problem (1,3,88,89,90-92).

Cutaneous Leishmaniasis

Choosing to treat will depend on the risk of (future) mucosal leishmaniasis, on the lesions' locations (joints, face) or number, whether or not joint mobility is hampered, the presence of secondary infection or of sporotrichoid, lymphangitic spread, etc. *L. major, L. tropica* and *L. mexicana* may not require treatment at all (1,4,5). Unfortunately, few well-designed studies have been performed. Also, variable regional responses to certain drugs are often noted, even within the same species (93). Systemic antimony, (short-course in certain regions), ABD, aminosidine, pentamidine, and IFN-γ, intra-lesional antimony (highly variable timing and dosing schedules), oral therapy (e.g. ketoconazole, itraconazole, allopurinol, dapsone), various non-pharmacological therapies (heat, irradiation, cryotherapy, electricity, surgical excision) are used (1,4-6,94-99). Lipid-associated formulations of amphotericin B may not be efficacious for CL, and the paucity of data precludes any recommendation at this point (1,4,100). *L. aethiopica* may not respond to antimony and pentamidine may be required (6).

Topically applied aminosidine can be effective for certain *Leishmania* species (primarily *L. major*) though irritation at the site of the ulcer can be intense, probably due to the irritant vehicle, methylbenzethonium chloride (101-103). Intra- and inter- specific variability of patient responses has been noted. Topical imidazoles have been disappointing (1,4-6). Pentamidine in low dose or for short duration may be useful and well tolerated in some new world CL (1,4,4,5).

Miltefosine, WR6026, other compounds as well as immunotherapy are also currently under investigation (104). DCL and leishmaniasis recidivans tend to respond poorly to therapy.

Mucosal Leishmaniasis

Usually a rare sequella of new-world cutaneous leishmaniasis (<5%), this rare manifestation of the *Viannia* subspecies might be prevented if the infection is treated early, when only cutaneous disease is evident. Clearly, even without evidence of metastatic disease, local therapy alone is unacceptable. Parenteral antimony is the mainstay of therapy, though amphotericin B, pentamidine, ketoconazole and recently itraconazole have been tried (1,4-6,105). Mucosal disease is usually more difficult to treat than cutaneous infection due to the same parasite (1,4,5). Novel therapies including a combination of drug and immunization for this entity are being studied.

A detailed summary of antileishmanial drugs and doses by clinical syndrome is presented in the appendix.

MODE OF ACTION AND MECHANISM OF RESISTANCE
Antimony
Sources

Organic pentavalent antimony in the form of sodium antimony gluconate has been used since the 1940's (106,107). Two such preparations, sodium stibogluconate and meglumine antimoniate have been used over three decades as first line VL chemotherapy (108). Current pentavalent antimony preparations include sodium stibogluconate, available as Pentostam (Wellcome Foundation Ltd., London) and meglumine antimoniate, available as Glucantime (Specia Rhone-Poulence Rorer, France and Rhodia Farma Ltd., Brazil) that is used mostly in South America. In addition, Indian pentavalent antimony is available in three locally produced forms, sodium antimony gluconate (Albert David Ltd., Calcutta), Stibanate (Gluconate India Pvt. Ltd., Calcutta) and Abanaye (Amoco Pharma, Muzaffarpur). A Chinese preparation (Shandong Xinhua, Pharmaceutical Factory) is also available (109).

Toxicity

Pentostam contains, in addition to sodium stibogluconate, the antiseptic (preservative?) m-chlorocresol, which is toxic to both promastigotes and amastigotes, and therefore potentially enhances Pentostam toxicity to *Leishmania* (110). When chlorocresol is omitted from Pentostam, toxicity to promastigotes is lost but toxicity to intracellular amastigotes remains. Furthermore, mutants of *L. donovani* promastigotes resistant to Pentostam are resistant to both chlorocresol and sodium stibogluconate (111). Sodium stibogluconate, when devoid of toxic preservatives, is less active against promastigotes than amastigotes in both infected macrophages and axenic culture (112,113). Meglumine antimoniate (Glucantime) is not marketed with toxic preservatives and displays identical stage specificity. The stage-specific toxicity responsible for the preferential susceptibility of intracellular amastigotes to Sb^{V} raises the question whether Sb^{V} is directly active against amastigotes or whether it is reduced by the host and/or parasite to Sb^{III}, which is highly toxic to both life stages (112). Studies using axenic host-free cultures indicate that amastigotes of *L. donovani* are >100 times more susceptible to Sb^{V} than promastigotes (113,114). Furthermore, in promastigotes Sb^{V} neutralizes the toxicity of Sb^{III} whereas both oxidation states additively kill axenic amastigotes in host-free culture (115). Interestingly, some axenic amastigotes of antimony resistant mutants resemble wild type promastigotes in that Sb^{V} neutralizes Sb^{III} toxicity (Gildor, T., Ephros, M., Ulrich, N. and Zilberstein, D., unpublished results). These observations suggest that pentavalent antimony is directly active against amastigotes. In contrast, recent studies in *L. infantum* showed that mutant amastigotes resistant to Sb^{V} in axenic culture are highly susceptible to both Sb oxidation states when inside macrophages. Furthermore, mutants resistant to Sb^{III} in culture are also resistant to both Sb^{III} and Sb^{V} in macrophages (116), suggesting that inside macrophages Sb^{III} and not Sb^{V} is responsible for parasite killing. Hence, SbV activity against *Leishmania* is still an open question.

Uptake and reduction

Sb^{III} and Sb^{V} enter *L. donovani* amastigotes and promastigotes via separate transport systems. Subsequently, a portion of the Sb^{V} is reduced to Sb^{III}, and both oxidation states form complexes with intracellular molecules (115). Although the nature of these intracellular interactions are not clear, their significance and correlation to parasite susceptibility is indicated by differences in the binding pattern of the Sb oxidation states to the intracellular molecules in promastigotes and amastigotes of wild type and Sb resistant mutants (115).

The mechanism of Sb^{V} reduction is unknown and the electron donor and enzyme mediating the reduction reaction have not been identified. Thiol containing compounds, e.g. glutathione and trypanothione (diglutathione spermidine), reduce Sb^{V} *in vitro* and it has been suggested they play a role in its toxicity (117). Glutaredoxin serves as the hydrogen donor for As^{V} reduction in *E. coli* and yeast (118,119). In staphylococci thioredoxin performs this function (120). Tryparedoxin, the glutaredoxin homologue of trypanosomatids may act as the electron donor in leishmanial Sb^{V} reduction (121). Tryparedoxin has been best characterized in the trypanosomatid *Crithidia fasciculate*. Its gene was cloned, and the structure of the protein was solved by X-ray crystallography (122-124). Tryparedoxin genes have been cloned in *T. cruzi* and *T. brucei* (125-127) and in *L. major* (128). However, the mechanism of Sb^V reduction by tryparedoxin has not yet been elucidated. Despite decades of use of Sb^V-containing compounds as anti-leishmanial agents, the mechanism of Sb^V reduction and its mode of action remain open questions.

Mechanism of resistance

L. tarentolae neutralizes As^{III} by complexing it with trypanothione coupled to either pumping of the complex out of the cells via an ATP-dependent pump or by sequestration of these complexes into an intracellular vesicle near the flagellar pocket (129-132). Trypanothione (diglutathione spermidine) is the major intracellular thiol in *Leishmania* (133). Its level increases in both Sb^{III} and As^{III} resistant mutants due to the amplification of *gsh1*, a gene that codes for γ glutamylcystein reductase (134). Moreover, in these mutants the level of *pgpa* that codes for P glycoprotein A, a protein that belongs to the family of ABC transporters, is also amplified in these mutants, further supporting the idea that Sb^{III} detoxification resembles that of As^{III} (135). Another feature of these mutants is that they contain high levels of polyamine and ornithine decarboxylase (136). Although the role of trypanothione and ornithine decarboxylase in metalloid resistance in *Leishmania* have been investigated only in promastigotes, it is likely that amastigotes possess similar mechanisms. These can now be addressed by using axenic amastigotes in host free systems (137-139). It is interesting to note that in all of the antimony and arsenite resistant mutants whose resistance is due to gene amplification, cross resistance between As^{III}, Sb^{III} and Sb^V is observed, whereas in the other cases, cross resistance is not obligatory (113,135,140).

MILTEFOSINE
Mode of action

Miltefosine (hexadecylphosphocholine) belongs to a family of alkyl- lysophospholipids, analogs of ether-lipids that inhibit sterol and phospholipid biosynthesis. This group of novel compounds have both *in vivo* and *in vitro* anti-neoplastic activity, pro-differentiation activity and can induce programmed cell death (141-146). Recent studies indicated that the anti-proliferative activity is not obtained via classical cytostatic effects, but rather by interfering with phosphinositol metabolism and signal transduction pathways. This group of ether-lipid analogs was originally developed as anti-neoplastic agents.

Miltefosine is an oral agent, which is highly active against *Leishmania* both *in vitro* and *in vivo*. It is also active against (clinically) antimony resistant visceral leishmnaiasis (due to *L.donovani*) (3). Recently, a topical formulation for *L. major* has been developed (Miltex) (147). The IC_{50} for promastigotes is approximately 17 μM and for intracellular amastigotes 0.2-5 μM in *L. donovani* and 16 μM in *L. tropica* (10,148). It is also active against axenic amastigotes with an IC_{50} of 1.5 μM {Sahked-Mishan, Ephros and Zilberstein unpublished data).

The abundance of alkyl-phosphlipids in *Leishmania* is likely the major reason for miltefosine's potent anti-leishmanial activity. Possible intracellular targets include alkyl-lipid metabolism and alkyl-anchored glycolipid (e.g. lipophosphoglycan) and glycoprotein (e.g. gp63) biosynthesis. Cytosolic enzymes involved in the initiation of ether lipid synthesis such as acetyl coenzyme A synthetase and acetyl coenzyme A reductase as well as the glycosomal dihydroxyacetone phosphate acetyltransferase are not affected by miltefosine. Later reactions that are also localized in glycosomes but are responsible for ether-lipid remodeling are affected: miltefosine inhibits alkyl-specific-acyl-coenzyme A acyltransferase in a dose dependent manner (149).

Mechanism of resistance

To date there have been no documented cases of miltefosine resistant parasites in kala azar patients. A limited number of laboratory studies predict that multi-drug resistance (e.g. P-glycoproteins) might modulate resistance to alkyl- lysophospholipids (150). Recent study by Perez-Victoria et al. indicate that overexpression of P glycoprotein (PGPA) in *L. tropica* induces a 9 and 7-fold resistance to miltefosine and edelfosine, respectively. There was a direct correlation between the level of P-glycoprotein expression and resistance to both drugs (148). More studies must be carried out to elucidate potential mechanisms of miltefosine resistance in *Leishmania*.

AMPHOTERICIN B
Mode of action

Amphotericin B is a polyene antibiotic used for more than 30 years for treating mycoses. It has selective activity against fungi, *Leishmania* and *T. cruzi* due to its greater affinity for the predominant microbial sterol, ergosterol, than for the host's cholesterol. Since 1990 it has been used as a second-line treatment for kala azar patients. It is highly effective in patients that are non-responsive to treatment with pentavalent antimony. Field studies indicate a long-term cure of over 90% of patients treated with amphotericin B (2).

In 1997, liposomal amphotericin B was licensed in the United States and other countries for use as anti leishmanial drug (9). Liposomal amphotericin B is more than 95 percent effective and is generally well tolerated; it and the other new lipid formulations of amphotericin B can be used for treating visceral leishmaniasis.

In vitro studies in promastigotes indicate differences in species susceptibility to amphotericin B that is related to variations in the ergosterol content of parasite membranes (151). However, no such differences have been observed in intracellular amastigotes, suggesting that variations in ergosterol content are less significant than in promastigotes.

Mechanism of resistance

One of the major advantages of amphotericin B over antimonials is the lack of resistance in the field. To date, no cases of amphotericin B resistance in kala azar patients have been reported. Furthermore, most of antimony resistant patients respond well to treatment with amphotericin B (3,152,153). Amphotericin B resistance has been studied in the laboratory strains. Resistant clones of *L. donovani* promastigotes have been raised by stepwise methods (154). Changes in sterol profile were observed in these clones: ergosterol is replaced with a precursor cholestra $-5,7,24$-trien 3β ol.

PENTAMIDINE
Mode of action

The aromatic diamidine pentamidine has been used to treat kala azar since around 1940, and despite its toxicity, pentamidine was initially viewed as effective and useful, but its efficacy declined rapidly (5,106,107,155). Pentamidine was the first alternative drug to be used in India and cured 99% of Sb^V non-responsive patients. However, over time even with double doses, it currently cures only 69-78% of patients and its use has largely been abandoned in India (156).

Pentamidine's mode of action has not been elucidated. Promastigotes take up pentamidine via polyamine transporter systems such as putrecine and spermidine (157-160). Several studies suggest that polyamine metabolism as well as mitochondrial functions are potential targets. Treatment of wild-type clones with low pentamidine concentrations for 24 hr provoke a strong decrease in arginine, ornithine, and putrescine pools, while the level of intracellular spermidine remains unchanged (161). Promastigotes resistant to pentamidine show changes in lipid membrane composition, mostly reduced levels of polar lipids (162). Similar pentamidine exposure also lowers(?) leishmanial mitochondrial membrane potential.

Mechanism of resistance

The mechanism of *Leishmania* resistance to pentamidine in the field has not been elucidated. Most studies on resistance have been carried out on a laboratory strains. Resistance involves reduced levels of pentamidine as well as polyamine transport (163). Mutant promastigotes resistant to pentamidine possess higher levels of putrescine than wild type, but lower pools of arginine and ornithine. Analysis by Western blot and DFMO-binding shows that resistant mutants have reduced amounts of ornithine decarboxylase (161).

CONCLUDING REMARKS

Drugs under development should ideally be continually available and orally administered, highly effective after a very short treatment course (with cure rates >95% in all geographic areas), against all species of parasite, in the old and young, during pregnancy and even in compromised hosts. They should also be safe/non-toxic and with no need for monitoring, acceptable, very cheap, and should be used either in combination with other anti-leishmanial agents (as with antituberculous, antileprosy, anti-HIV and antineoplastic therapeutic protocols) or simply be unaffected by parasite resistance mechanisms.

The future will include new and "engineered" compounds, better targeted against the parasites, once the molecular pathogenesis of the leishmaniases will be elucidated (72,164,165). Multiple drug regimens will probably be the rule and will include immunomodulators (166).

REFERENCES

1.Herwaldt, B. L. 1999. Leishmaniasis. Lancet 354: 1191-1199.
2. Murray, H. W. 2001. Clinical and experimental advances in treatment of visceral leishmaniasis. Antimicrob.Agents Chemother. 45: 2185-2197
3.Murray, H. W. 2000. Treatment of visceral leishmaniasis kala-azar: a decade of progress and future approaches. Int.J.Infect.Dis. 4: 158-177.
4.Moskowitz, P. F. and Kurban, A. K. 1999. Treatment of cutaneous leishmaniasis: retrospectives and advances for the 21st century. Clin.Dermatol. 17: 305-315. 5.Berman, J. D. 1997. Human leishmaniasis: clinical, diagnostic, and chemotherapeutic developments in the last 10 years. Clin.Infect.Dis. 24: 684-703.
6. Diseases, Principles, Pathogens, and Practice" R. L. Guerrant, D. H. Walker, and P. F. Weller, Eds., Churchill Livingstone, Philadelphia.
7.Davidson, R. N., Di Martino, L., Gradoni, L., Giacchino, R., Gaeta, G. B., Pempinello, R., Scotti, S., Cascio, A., Castagnola, E., Maisto, A., Gramiccia, M., di Caprio, D., Wilkinson, R. J., and Bryceson, A. D. 1996. Short-course treatment of visceral leishmaniasis with liposomal amphotericin B AmBisome. Clin.Infect.Dis. 22: 938-943. Meyerhoff, A. 1999.
8.US Food and Drug Administration approval of AmBisome liposomal amphotericin B for treatment of visceral leishmaniasis. Clin.Infect.Dis. 28: 42-48.
9. liposomal amphotericin B for treatment of visceral leishmaniasis - Editorial response. Clin.Infect.Dis. 28, 49-51.
10.Croft, S. L., Snowdon, D., and Yardley, V. 1996. The activities of four anticancer alkyllysophospholipids against *Leishmania donovani*, *Trypanosoma cruzi* and *Trypanosoma brucei*. J.Antimicrob.Chemother. 38: 1041-1047.
11.Kuhlencord, A., Maniera, T., Eibl, H., and Unger, C. 1992. Hexadecylphosphocholine: oral treatment of visceral leishmaniasis in mice. Antimicrob.Agents Chemother. 36: 1630-1634.

12.Berman, J. D., Badaro, R., Thakur, C. P., Wasunna, K. M., Behbehani, K., Davidson, R., Kuzoe, F., Pang, L., Weerasuriya, K., and Bryceson, A. D. 1998. Efficacy and safety of liposomal amphotericin B AmBisome for visceral leishmaniasis in endemic developing countries. Bull.World Health Organ 76: 25-32.

13.Dietze, R., Fagundes, S. M., Brito, E. F., Milan, E. P., Feitosa, T. F., Suassuna, F. A., Fonschiffrey, G., Ksionski, G., and Dember, J. 1995. Treatment of kala-azar in Brazil with Amphocil amphotericin B cholesterol dispersion for 5 days. Trans.R.Soc.Trop.Med.Hyg. 89: 309-311.

14. Di Martino, L., Davidson, R. N., Giacchino, R., Scotti, S., Raimondi, F., Castagnola, E., Tasso, L., Cascio, A., Gradoni, L., Gramiccia, M., Pettoello-Mantovani, M., and Bryceson, A. D. 1997. Treatment of visceral leishmaniasis in children with liposomal amphotericin B. J.Pediatr. 131: 271-277.

15.Sundar, S. and Murray, H. W. 1996. Cure of antimony-unresponsive Indian visceral leishmaniasis with amphotericin B lipid complex. J.Infect.Dis. 173: 762-765.

16. Sundar, S., Agrawal, N. K., Sinha, P. R., Horwith, G. S., and Murray, H. W. 1997. Short-course, low-dose amphotericin B lipid complex therapy for visceral leishmaniasis unresponsive to antimony. Ann.Intern.Med. 127: 133-137.

17.Seaman, J., Boer, C., Wilkinson, R., de Jong, J., de Wilde, E., Sondorp, E., and Davidson, R. 1995. Liposomal amphotericin B AmBisome in the treatment of complicated kala-azar under field conditions. Clin.Infect.Dis. 21: 188-193.

18.Croft, S. L., Neal, R. A., Pendergast, W., and Chan, J. H. 1987. The activity of alkyl phosphorylcholines and related derivatives against *Leishmania donovani*. Biochem.Pharmacol. 36: 2633-2636.

19. Le Fichoux, Y., Rousseau, D., Ferrua, B., Ruette, S., Lelievre, A., Grousson, D., and Kubar, J. 1998. Short- and jong-term efficacy of hexadecylphosphocholine against *Leishmania infantum* infection in BALB/c mice. Antimicrob.Agents Chemother. 42: 654 658.

20.Sundar, S., Rosenkaimer, F., Makharia, M. K., Goyal, A. K., Mandal, A. K., Voss, A., Hilgard, P., and Murray, H. W. 1998. Trial of oral miltefosine for visceral leishmaniasis. Lancet 352: 1821-1823.

21. Sundar, S., Gupta, L. B., Makharia, M. K., Singh, M. K., Voss, A., Rosenkaimer, F., Engel, J., and Murray, H. W. 1999. Oral treatment of visceral leishmaniasis with miltefosine. Ann.Trop.Med.Parasitol. 93: 589-597.

22. Sundar, S., Makharia, A., More, D. K., Agrawal, G., Voss, A., Fischer, C., Bachmann, P., and Murray, H. W. 2000. Short-course of oral miltefosine for treatment of visceral leishmaniasis. Clin.Infect.Dis. 31: 1110-1113.

23. Croft, S. L., Neal, R. A., Thornton, E. A., and Herrmann, D. B. 1993. Antileishmanial activity of the ether phospholipid ilmofosine. Trans.R.Soc.Trop.Med.Hyg. 87: 217-219.

24. Chance, M. L. 1995. New developments in the chemotherapy of leishmaniasis. Ann.Trop.Med.Parasitol. 89:37-43:37-43.

25. Ouellette, M. 2001. Biochemical and molecular mechanisms of drug resistance in parasites. Trop Med Int.Health 6: 874-882.

26. Ouellette, M., Legare, D., Haimeur, A., Grondin, K., Ray, G., Brochu, C., and Papadopoulou, B. 1998. ABC transporters in *Leishmania* and their role in drug resistance. Drug resistance Updates 1: 43-48.

27. Ouellette, M., Legare, D., and Papadopoulou, B. 2001. Multidrug resistance and ABC transporters in parasitic protozoa. J.Mol.Microbiol.Biotechnol. 3: 201-206.

28. Ullmann, B. 1995. Multidrug resistance and P-glycoprotein in parasitic protozoa. J.Bioenerg.Biomembr. 27: 77-84.

29. Beverley, S. M. 1991. Gene amplification in *Leishmania*. Annu.Rev.Microbiol. 45: 417-444.

30. Chow, C. M. and Volkman, S. K. 1998. Plasmodium and *Leishmania*: the role of

mdr genes in mediating drug resistance. Exp.Parasitol 90: 135-141.

31. Pampiglione, S., Manson Bahr, P. E., La Placa, M., Borgatti, M. A., and Musumeci, S. 1975. Studies in Mediterranean leishmaniasis. 3. The leishmanin skin test in kala azar. Trans.R.Soc.Trop.Med.Hyg. 69: 60-68.

32. Nabors, G. S. and Farrell, J. P. 1994. Depletion of interleukin-4 in BALB/c mice with established *Leishmania major* infections increases the efficacy of antimony therapy and promotes Th1-like responses. Infect.Immun. 62: 5498-5504.

33. Meller-Melloul, C., Farnarier, C., Dunan, S., Faugere, B., Franck, J., Mary, C., Bongrand, P., Quilici, M., and Kaplanski, S. 1991. Evidence of subjects sensitized to *Leishmania infantum* on the French Mediterranean coast: differences in gamma interferon production between this population and visceral leishmaniasis patients. Parasite Immunol. 13: 531-536.

34. Sacks, D. L., Lal, S. L., Shrivastava, S. N., Blackwell, J., and Neva, F. A. 1987. An analysis of T cell responsiveness in Indian kala-azar. J.Immunol. 138: 908-913.

35. Carvalho, E. M., Barral, A., Pedral-Sampaio, D., Barral-Netto, M., Badaro, R., Rocha, H., and Johnson, W. D., Jr. 1992. Immunologic markers of clinical evolution in children recently infected with *Leishmania donovani* chagasi. J.Infect.Dis. 165: 535-540.

36. D'Oliveira, J. A., Costa, S. R., Barbosa, A. B., Orge, M. d. l., and Carvalho, E. M. 1997. Asymptomatic *Leishmania* chagasi infection in relatives and neighbors of patients with visceral leishmaniasis. Mem.Inst.Oswaldo Cruz 92: 15-20.

37. Saran, R., Gupta, A. K., and Sharma, M. C. 1991. Leishmanin skin test in clinical and subclinical kala-azar cases. J.Commun.Dis. 23: 135-137.

38. Murray, H. W. and Delph-Etienne, S. 2000. Roles of endogenous gamma interferon and macrophage microbicidal mechanisms in host response to chemotherapy in experimental visceral leishmaniasis. Infect.Immun. 68: 288-293.

39. Kaye, P. M., Curry, A. J., and Blackwell, J. M. 1991. Differential production of Th1- and Th2-derived cytokines does not determine the genetically controlled or vaccine-induced rate of cure in murine visceral leishmaniasis. J.Immunol. 146: 2763-2770.

40. Scharton-Kersten, T., Afonso, L. C., Wysocka, M., Trinchieri, G., and Scott, P. 1995. IL-12 is required for natural killer cell activation and subsequent T helper 1 cell development in experimental leishmaniasis. J.Immunol. 154: 5320-5330.

41. Reiner, S. L. and Locksley, R. M. 1995. The regulation of immunity to *Leishmania major*. Annu.Rev.Immunol. 13: 151-177.

42. Sang, D. K., Ouma, J. H., John, C. C., Whalen, C. C., King, C. L., Mahmoud, A. A., and Heinzel, F. P. 1999. Increased levels of soluble interleukin-4 receptor in the sera of patients with visceral leishmaniasis. J.Infect.Dis. 179: 743-746.

43. Ghalib, H. W., Piuvezam, M. R., Skeiky, Y. A., Siddig, M., Hashim, F. A., El Hassan, A. M., Russo, D. M., and Reed, S. G. 1993. Interleukin 10 production correlates with pathology in human *Leishmania donovani* infections. J.Clin.Invest 92: 324-329.

44. Murray, H. W., Hariprashad, J., and Coffman, R. L. 1997. Behavior of visceral *Leishmania donovani* in an experimentally induced T helper cell 2 Th2-associated response model. J.Exp.Med. 185: 867-874.

45. Barral-Netto, M., Barral, A., Brownell, C. E., Skeiky, Y. A., Ellingsworth, L. R., Twardzik, D. R., and Reed, S. G. 1992. Transforming growth factor-beta in *leishmanial* infection: a parasite escape mechanism. Science 257: 545-548.

46. Wilson, M. E., Young, B. M., Davidson, B. L., Mente, K. A., and McGowan, S. E. 1998. The importance of TGF-beta in murine visceral leishmaniasis. J.Immunol. 161: 6148-6155.

47. Sundar, S., Reed, S. G., Sharma, S., Mehrotra, A., and Murray, H. W. 1997. Circulating T helper 1 Th1 cell- and Th2 cell-associated cytokines in Indian patients with visceral leishmaniasis. Am.J.Trop.Med.Hyg. 56: 522-525.

48. Magill, A. J., Grogl, M., Gasser, R. A., Jr., Sun, W., and Oster, C. N. 1993. Visceral infection caused by *Leishmania tropica* in veterans of Operation Desert Storm. N.Engl.J.Med. 328: 1383-1387.

49. Magill, A. J., Grogl, M., Johnson, S. C., and Gasser, R. A., Jr. 1994. Visceral infection due to *Leishmania tropica* in a veteran of Operation Desert Storm who presented 2 years after leaving Saudi Arabia. Clin.Infect.Dis. 19: 805-806.

50. Alvar, J., Canavate, C., Gutierrez-Solar, B., Jimenez, M., Laguna, F., and Lopez Velez, R. 1997. *Leishmania* and human immunodeficiency virus coinfection: the first 10 years. Clininical Microbiology Review 10: 298-319.

51. Sundar, S., More, D. K., Singh, M. K., Singh, V. P., Sharma, S., Makharia, A., Kumar, P. C., and Murray, H. W. 2000. Failure of pentavalent antimony in visceral leishmaniasis in India: report from the center of the Indian epidemic. Clin.Infect.Dis. 31: 1104-1107.

52. Murray, H. W., Hariprashad, J., and Fichtl, R. E. 1993. Treatment of experimental visceral leishmaniasis in a T-cell- deficient host: response to amphotericin B and pentamidine. Antimicrob.Agents Chemother. 37: 1504-1505.

53. Ribera, E., Ocana, I., de Otero, J., Cortes, E., Gasser, I., and Pahissa, A. 1996. Prophylaxis of visceral leishmaniasis in human immunodeficiency virus- infected patients. Am.J.Med. 100: 496-501.

54. Desjeux, P., Piot, B., O'Neill, K., and Meert, J. P. 2001. Co-infections of *leishmania*/HIV in south Europe. Med.Trop.Mars. 61: 187-193.

55. Montalban, C., Calleja, J. L., Erice, A., Laguna, F., Clotet, B., Podzamczer, D., Cobo, J., Mallolas, J., Yebra, M., and Gallego, A. 1990. Visceral leishmaniasis in patients infected with human immunodeficiency virus. Co-operative Group for the Study of Leishmaniasis in AIDS. J.Infect. 21: 261-270.

56. Laguna, F., Lopez-Velez, R., Pulido, F., Salas, A., Torre-Cisneros, J., Torres, E., Medrano, F. J., Sanz, J., Pico, G., Gomez-Rodrigo, J., Pasquau, J., and Alvar, J. 1999. Treatment of visceral leishmaniasis in HIV-infected patients: a randomized trial comparing meglumine antimoniate with amphotericin B. Spanish HIV-Leishmania Study Group. AIDS 13: 1063-1069.

57. Wali, J. P., Aggarwal, P., Nandy, A., Singh, S., Addy, M., Guha, S. K., Dwivedi, S. N., Karmarkar, M. G., and Maji, A. K. 1997. Efficacy of sodium antimony gluconate and ketoconazole in the treatment of kala-azar--a comparative study. J.Commun.Dis. 29: 73-83.

58. Sherwood JA, Gachihi GS, Muigai RK, Skillman DR, Mugo M, Rashid JR, Wasunna KM, Were JB, Kasili SK, Mbugua JM, et al. 1994. Phase 2 efficacy trial of an oral 8 aminoquinoline WR6026 for treatment of visceral leishmaniasis. Clin Infect Dis. 19: 1034-1039.

59. Sundar, S., Singh, V. P., Agrawal, N. K., Gibbs, D. L., and Murray, H. W. 1996. Treatment of kala-azar with oral fluconazole. Lancet 348: 614.

60. Sundar, S., Kumar, P., Makharia, M., Goyal, A., Rogers, M., Gibbs, D., and Murray, H. 1998. Atovaquone alone or with fluconazole as oral therapy for Indian kala- azar. Clin.Infect.Dis. 27v 215-216.

61. Sherwood, J. A., Gachihi, G. S., Muigai, R. K., Skillman, D. R., Mugo, M., Rashid, J. R., Wasunna, K. M., Were, J. B., Kasili, S. K., Mbugua, J. M., and . 1994. Phase 2 efficacy trial of an oral 8-aminoquinoline WR6026 for treatment of visceral leishmaniasis. Clin.Infect.Dis. 19: 1034-1039.

62. Seaman, J., Pryce, D., Sondorp, H. E., Moody, A., Bryceson, A. D., and Davidson, R. N. 1993. Epidemic visceral leishmaniasis in Sudan: a randomized trial of aminosidine plus sodium stibogluconate versus sodium stibogluconate alone. J.Infect.Dis. 168: 715-720.

63. Thakur, C. P., Kanyok, T. P., Pandey, A. K., Sinha, G. P., Zaniewski, A. E., Houlihan, H. H., and Olliaro, P. 2000. A prospective randomized, comparative, open

label trial of the safety and efficacy of paromomycin aminosidine plus sodium stibogluconate versus sodium stibogluconate alone for the treatment of visceral leishmaniasis. Trans.R.Soc.Trop.Med.Hyg. 94: 429-431.

64. Sundar, S., Rosenkaimer, F., and Murray, H. W. 1994. Successful treatment of refractory visceral leishmaniasis in India using antimony plus interferon-gamma. J.Infect.Dis. 170: 659-662.

65. Badaro, R., Nascimento, C., Carvalho, J. S., Badaro, F., Russo, D., Ho, J. L., Reed, S. G., Johnson, W. D., Jr., and Jones, T. C. 1994. Recombinant human granulocyte macrophage colony-stimulating factor reverses neutropenia and reduces secondary nfections in visceral leishmaniasis. J.Infect.Dis. 170: 413-418.

66. Badaro, R., Falcoff, E., Badaro, F. S., Carvalho, E. M., Pedral-Sampaio, D., Barral, A., Carvalho, J. S., Barral-Netto, M., Brandely, M., Silva, L., and . 1990. Treatment of visceral leishmaniasis with pentavalent antimony and interferon gamma. N.Engl.J.Med. 322: 16-21.

67. Sundar, S., Rosenkaimer, F., Lesser, M. L., and Murray, H. W. 1995. Immunochemotherapy for a systemic intracellular infection: accelerated response using interferon-gamma in visceral leishmaniasis. J.Infect.Dis. 171: 992-996.

68. Murray, H. W., Cervia, J. S., Hariprashad, J., Taylor, A. P., Stoeckle, M. Y., and Hockman, H. 1995. Effect of granulocyte-macrophage colony-stimulating factor in experimental visceral leishmaniasis. J.Clin.Invest 95: 1183-1192.

69. Almeida, R., D'Oliveira, A., Jr., Machado, P., Bacellar, O., Ko, A. I., de Jesus, A. R., Mobashery, N., Brito, S. J., and Carvalho, E. M. 1999. Randomized, double-blind study of stibogluconate plus human granulocyte macrophage colony-stimulating factor versus stibogluconate alone in the treatment of cutaneous leishmaniasis. J.Infect.Dis. 180: 1735-1737.

70. Herwaldt, B. L. and Berman, J. D. 1992. Recommendations for treating leishmaniasis with sodium stibogluconate Pentostam and review of pertinent clinical studies. Am.J.Trop.Med.Hyg. 46: 296-306.

71. Aronson, N. E., Wortmann, G. W., Johnson, S. C., Jackson, J. E., Gasser, R. A., Jr., Magill, A. J., Endy, T. P., Coyne, P. E., Grogl, M., Benson, P. M., Beard, J. S., Tally, J. D., Gambel, J. M., Kreutzer, R. D., and Oster, C. N. 1998. Safety and efficacy of intravenous sodium stibogluconate in the treatment of leishmaniasis: recent U.S. military experience. Clin.Infect.Dis. 27: 1457-1464.

72. Agrawal, A. K. and Gupta, C. M. 2000. Tuftsin-bearing liposomes in treatment of macrophage-based infections. Adv.Drug Deliv.Rev. 41: 135-146.

73. Bora, D. 1999. Epidemiology of visceral leishmaniasis in India. Natl.Med.J.India. 12: 62-68.

74. Lira, R., Sundar, S., Makharia, A., Kenney, R., Gam, A., Saraiva, E., and Sacks, D. 1999. Evidence that the high incidence of treatment failures in Indian kala- azar is due to the emergence of antimony-resistant strains of Leishmania donovani. J.Infect.Dis. 180: 564-567.

75. Khalil, E. A., El Hassan, A. M., Zijlstra, E. E., Hashim, F. A., Ibrahim, M. E., Ghalib, H. W., and Ali, M. S. 1998. Treatment of visceral leishmaniasis with sodium stibogluconate in Sudan: management of those who do not respond. Ann.Trop Med Parasitol. 92: 151-158.

76. Seaman, J., Mercer, A. J., Sondorp, H. E., and Herwaldt, B. L. 1996. Epidemic visceral leishmaniasis in southern Sudan: treatment of severely debilitated patients under wartime conditions and with limited resources. Ann.Intern.Med. 124: 664-672.

77. Sundar, S., Thakur, B. B., Tandon, A. K., Agrawal, N. R., Mishra, C. P., Mahapatra, T. M., and Singh, V. P. 1994. Clinicoepidemiological study of drug resistance in Indian kala-azar. BMJ 308: 307.

78. Thakur, C. P., Sinha, G. P., Pandey, A. K., Kumar, N., Kumar, P., Hassan, S. M., Narain, S., and Roy, R. K. 1998. Do the diminishing efficacy and increasing toxicity of

sodium stibogluconate in the treatment of visceral leishmaniasis in Bihar, India, justify its continued use as a first-line drug? An observational study of 80 cases. Ann.Trop.Med.Parasitol. 92: 561-569.

79. Sievers, T. M., Kubak, B. M., and Wong-Beringer, A. 1996. Safety and efficacy of intralipid emulsions of amphotericin B. J.Antimicrob.Chemother. 38: 333-347.

80. Sundar, S., Gupta, L. B., Rastogi, V., Agrawal, G., and Murray, H. W. 2000. Short course, cost-effective treatment with amphotericin B-fat emulsion cures visceral leishmaniasis. Trans.R.Soc.Trop.Med.Hyg. 94: 200-204.

81. Bodhe, P. V., Kotwani, R. N., Kirodian, B. G., Pathare, A. V., Pandey, A. K., Thakur, C. P., and Kshirsagar, N. A. 1999. Dose-ranging studies on liposomal amphotericin B L-AMP-LRC-1 in the treatment of visceral leishmaniasis. Trans.R.Soc.Trop.Med.Hyg. 93: 314-318.

82. Jha, T. K., Olliaro, P., Thakur, C. P., Kanyok, T. P., Singhania, B. L., Singh, I. J., Singh, N. K., Akhoury, S., and Jha, S. 1998. Randomised controlled trial of aminosidine paromomycin v sodium stibogluconate for treating visceral leishmaniasis in North Bihar, India. BMJ 316: 1200-1205.

83. Thakur, C. P., Kanyok, T. P., Pandey, A. K., Sinha, G. P., Messick, C., and Olliaro, P. 2000. Treatment of visceral leishmaniasis with injectable paromomycin aminosidine. An open-label randomized phase-II clinical study. Trans.R.Soc.Trop.Med.Hyg. 94: 432-433.

84. Berman, J. D. and Lee, L. S. 1984. Activity of antileishmanial agents against amastigotes in human monocyte-derived macrophages and in mouse peritoneal macrophages. J.Parasitol. 70: 220-225.

85. Neal, R. A. and Croft, S. L. 1984. An in-vitro system for determining the activity of compounds against the intracellular amastigote form of *Leishmania donovani*. J.Antimicrob.Chemother. 14: 463-475.

86. Berman, J. D. and Lee, L. S. 1983. Activity of 8-aminoquinolines against *Leishmania tropica* within human macrophages in vitro. Am.J.Trop.Med.Hyg. 32: 753 759.

87. Bitnun, A., Giladi, M., and Efrat, M. 1998. Leishmaniasis in Pregnancy. In "Textbook of Perinatal Medicine" A. Kurjak, Ed., Parthenon Publishing, London.

88. Ribera, E., Ocana, I., de Otero, J., Cortes, E., Gasser, I., and Pahissa, A. 1996. Prophylaxis of visceral leishmaniasis in human immunodeficiency virus- infected patients. Am.J.Med. 100: 496-501.

89. Murray, H. W. 1999. Kala-azar as an AIDS-related opportunistic infection. AIDS Patient.Care STDS. 13: 459-465.

90. Alvar, J., Canavate, C., Gutierrez-Solar, B., Jimenez, M., Laguna, F., Lopez-Velez, R., Molina, R., and Moreno, J. 1997. Leishmania and human immunodeficiency virus co-infection: the first 10 years. Clin Microbiol.Rev. 10: 298-319.

91. Soriano, V., Dona, C., Rodriguez-Rosado, R., Barreiro, P., and Gonzalez-Lahoz, J. 2000. Discontinuation of secondary prophylaxis for opportunistic infections in HIV infected patients receiving highly active antiretroviral therapy. AIDS 14: 383-386.

92. Rosenthal, E., Tempesta, S., Del Giudice, P., Marty, P., Desjeux, P., Pradier, C., Le Fichoux, Y., and Cassuto, J. P. 2001. Declining incidence of visceral leishmaniasis in HIV-infected individuals in the era of highly active antiretroviral therapy. AIDS 15: 1184-1185.

93. Neva, F. A., Ponce, C., Ponce, E., Kreutzer, R., Modabber, F., and Olliaro, P. 1997. Non-ulcerative cutaneous leishmaniasis in Honduras fails to respond to topical paromomycin. Trans R Soc Trop Med Hyg. 91: 473-475.

94. Aste, N., Pau, M., Ferreli, C., and Biggio, P. 1998. Intralesional treatment of cutaneous leishmaniasis with meglumine antimoniate. Br.J.Dermatol. 138: 370-371.

95. Osorio, L. E., Palacios, R., Chica, M. E., and Ochoa, M. T. 1998. Treatment of cutaneous leishmaniasis in Colombia with dapsone. Lancet 351: 498-499.

96. Alkhawajah, A. M., Larbi, E., Al Gindan, Y., Abahussein, A., and Jain, S. 1997. Treatment of cutaneous leishmaniasis with antimony: intramuscular versus intralesional administration. Ann.Trop Med Parasitol. 91: 899-905.

97. Oliveira-Neto, M. P., Schubach, A., Mattos, M., Goncalves-Costa, S. C., and Pirmez, C. 1997. A low-dose antimony treatment in 159 patients with American cutaneous leishmaniasis: extensive follow-up studies up to 10 years. Am.J.Trop Med Hyg. 57: 651-655.

98. Oliveira-Neto, M. P., Schubach, A., Mattos, M., da Costa, S. C., and Pirmez, C. 1997. Intralesional therapy of American cutaneous leishmaniasis with pentavalent antimony in Rio de Janeiro, Brazil--an area of *Leishmania* V. braziliensis transmission. Int.J.Dermatol. 36: 463-468.

99. Nacher, M., Carme, B., Sainte, M. D., Couppie, P., Clyti, E., Guibert, P., and Pradinaud, R. 2001. Influence of clinical presentation on the efficacy of a short course of pentamidine in the treatment of cutaneous leishmaniasis in French Guiana. Ann.Trop Med Parasitol. 95: 331-336.

100. Wortmann, G. W., Fraser, S. L., Aronson, N. E., Davis, C., Miller, R. S., Jackson, J. D., and Oster, C. N. 1998. Failure of amphotericin B lipid complex in the treatment of cutaneous leishmaniasis. Clin Infect Dis. 26: 1006-1007.

101. el On, J., Livshin, R., Even Paz, Z., Hamburger, D., and Weinrauch, L. 1986. Topical treatment of cutaneous leishmaniasis. J.Invest.Dermatol. 87: 284-288.

102. el On, J., Halevy, S., Grunwald, M. H., and Weinrauch, L. 1992. Topical treatment of Old World cutaneous leishmaniasis caused by *Leishmania major*: a double blind control study. J.Am.Acad.Dermatol. 27: 227-231.

103. Ozgoztasi, O. and Baydar, I. 1997. A randomized clinical trial of topical paromomycin versus oral ketoconazole for treating cutaneous leishmaniasis in Turkey. Int.J.Dermatol. 36: 61-63.

104. Soto, J., Toledo, J., Gutierrez, P., Nicholls, R.S., Padilla, J., Engel, J., Fischer, C., Voss, A. and Berman, D 2001. Treatment of American cutaneous leishmaniasis with miltefosine, an oral agent. Clin.Infect.Dis. 33: E57-61.

105. Amato, V. S., Padilha, A. R., Nicodemo, A. C., Duarte, M. I., Valentini, M., Uip, D. E., Boulos, M., and Neto, V. A. 2000. Use of itraconazole in the treatment of mucocutaneous leishmaniasis: a pilot study. Int.J.Infect Dis. 4: 153-157.

106. Manson Bahr, P. E. C. 1959. East Africa kala-azar with special reference to the pathology prophylaxis, and treatment. Trans R Soc Trop Med Hyg. 53: 123-137.

107. Shortt, H. E. 1945. Recent research on kala-azar in India. Trans R Soc Trop Med Hyg. 39, 13-41.

108. Gupta, P. C. S. 1953. Chemotherapy of leishmanial diseases: a resume of recent researches. Indian Med.Gaz. 88: 20-35.

109. deFigueiredo, E. M., Silva, J. C., and Brazil, R. P. 1999. Experimental treatment with sodium stibogluconate of hamster infected with *Leishmania Leishmania chagasi* and *Leishmania Leishmania amazonensis*. Rev.Soc.Bras.Med.Trop. 32: 191-193.

110. Roberts, W. L. and Rainey, P. M. 1993. Antileishmanial activity of sodium stibogluconate fractions. Antimicrob.Agents Chemother. 37: 1842-1846.

111. Ephros, M., Waldman, E., and Zilberstein, D. 1997. Pentostam induces resistance to antimony and preservative chlorocresol in *Leishmania donovani* promastigotes and axenically grown amastigotes. Antimicrob.Agents Chemother. 41: 1064-1068.

112. Roberts, W. L., Berman, J. D., and Rainey, P. M. 1995. In vitro antileishmanial properties of tri- and pentavalent antimonial preparations. Antimicrob.AgentsChemother. 39: 1234-1239.

113. Ephros, M., Bitnun, A., Shaked, P., Waldman, E., and Zilberstein, D. 1999. Stage specific activity of pentavalent antimony against *Leishmania donovani* axenic amastigotes. Antimicrob.Agents Chemother. 43: 278-282.

114. Sereno, D. and Lemesre, J. L. 1997. Axenically cultured amastigote forms as an in

vitro model for investigation of antileishmanial agents. Antimicrob.Agents Chemother. 41: 972-976.

115. Shaked-Mishan, P., Ulrich.N., Ephros, M., and Zilberstein, D. 2001. Novel intracellular Sb^V reducing activity correlates with antimony susceptibility in *Leishmania donovani*. J.Biol.Chem. 276: 3971-3976.

116. Sereno, D., Cavaleyra, M., Zemzoumi, K., Maquaire, S., Ouaissi, A., and Lemesre, J. L. 1998. Axenically Growth Amastigotes of *Leishmania infantum* Used as an In Vitro Model To Investigate the Pentavalent Antimony Mode of Action. Antimicrob.Agents Chemother. 42: 3097-3102.

117. Frezard, F., Demicheli, C., Ferreira, C. S., and Costa, M. A. 2001. Glutathione induced conversion of pentavalent antimony to trivalent antimony in meglumine antimoniate. Antimicrob.Agents Chemother. 45: 913-916.

118. Mukhopadhyay, R., Shi, J., and Rosen, B. P. 2000. Purification and characterization of ACR2p, the Saccharomyces cerevisiae arsenate reductase. J.Biol.Chem. 275: 21149-21157.

119. Shi, J., Vlamis-Gardikas, A., Aslund, F., Holmgren, A., and Rosen, B. P. 1999. Reactivity of glutaredoxins 1, 2, and 3 from Escherichia coli shows that glutaredoxin 2 is the primary hydrogen donor to ArsC-catalyzed arsenate reduction. J.Biol.Chem. 274: 36039-36042.

120. Messens, J., Hayburn, G., Desmyter, A., Laus, G., and Wyns, L. 1999. The essential catalytic redox couple in arsenate reductase from Staphylococcus aureus. Biochemistry 38: 16857-16865.

121. Montemartini, M., Nogoceke, E., Gommel, D. U., Singh, M., Kalisz, H. M., Steinert, P., and Flohe, L. 2000. Tryparedoxin and tryparedoxin peroxidase. Biofactors 11: 71-72.

122. Tetaud, E. and Fairlamb, A. H. 1998. Cloning, expression and reconstitution of the trypanothione-dependent peroxidase system of Crithidia fasciculata. Mol.Biochem.Parasitol. 96: 111-123.

123. Guerrero, S. A., Montemartini, M., Spallek, R., Hecht, H. J., Steinert, P., Flohe, L., and Singh, M. 2000. Cloning and expression of tryparedoxin I from Crithidia fasciculata. Biofactors 11: 67-69.

124. Kalisz, H. M., Hofmann, B., Nogoceke, E., Gommel, D. U., Flohe, L., and Hecht, H. J. 2000. Crystallisation of tryparedoxin I from *Crithidia fasciculata*. Biofactors 11: 73-75.

125. Tetaud, E., Giroud, C., Prescott, A. R., Parkin, D. W., Baltz, D., Biteau, N., Baltz, T., and Fairlamb, A. H. 2001. Molecular characterisation of mitochondrial and cytosolic trypanothione- dependent tryparedoxin peroxidases in Trypanosoma brucei. Mol.Biochem.Parasitol. 116: 171-183.

126. Guerrero, S. A., Lopez, J. A., Steinert, P., Montemartini, M., Kalisz, H. M., Colli, W., Singh, M., Alves, M. J., and Flohe, L. 2000. His-tagged tryparedoxin peroxidase of Trypanosoma cruzi as a tool for drug screening. Appl.Microbiol.Biotechnol. 53: 410-414.

127. Avila, A. R., Yamada-Ogatta, S. F., da, S. M., V, Krieger, M. A., Nakamura, C. V., De Souza, W., and Goldenberg, S. 2001. Cloning and characterization of the metacyclogenin gene, which is specifically expressed during Trypanosoma cruzi metacyclogenesis. Mol.Biochem.Parasitol. 117: 169-177.

128. Levick, M. P., Tetaud, E., Fairlamb, A. H., and Blackwell, J. M. 1998. Identification and characterisation of a functional peroxidoxin from *Leishmania major*. Mol.Biochem.Parasitol 96: 125-137.

129. Dey, S., Papadopoulou, B., Roy, G., Grondin, K., Dou, D., Rosen, B. P., Ouellette, M., and Haimeur, A. 1994. High level arsenite resistance in *Leishmania tarentolae* is mediated by active extrusion system. Mol.Biochem.Parasitol 67: 49-57.

130. McKean, P. G., Keen, J. K., Smith, D. F., and Benson, F. E. 2001. Identification

and characterisation of a RAD51 gene from *Leishmania major*. Mol.Biochem.Parasitol. 115: 209-216.

131. Legare, D., Papadopoulou, B., Roy, G., Mukhopadhyay, R., Haimeur, A., Dey, S., Grondin, K., Brochu, C., Rosen, B. P., and Ouellette, M. 1997. Efflux systems and increased trypanothione levels in arsenite-resistant *Leishmania*. Exp.Parasitol. 87: 275 282.

132. Legare, D., Richard, D., Mukhopadhyay, R., Stierhof, Y. D., Rosen, B. P., Haimeur, A., Papadopoulou, B., and Ouellette, M. 2001. The *Leishmania* ABC protein PGPA is an intracellular metal-thiol transporter ATPase. J.Biol.Chem. 276:26301-26307

133. Ariyanayagam, M. R. and Fairlamb, A. H. 2001. Ovothiol and trypanothione as antioxidants in trypanosomatids. Mol.Biochem.Parasitol. 115: 189-198.

134. Grondin, K., Haimeur, A., Mukhopadhyay, R., Rosen, B. P., and Ouellette, M. 1997. Co-amplification of the gamma-glutamylcysteine synthetase gene gsh1 and of the ABC transporter gene pgpA in arsenite-resistant *Leishmania tarentolae*. EMBO J 16: 3057-3065.

135. Haimeur, A., Brochu, C., Genest, P., Papadopoulou, B., and Ouellette, M. 2000. Amplification of the ABC transporter gene PGPA and increased trypanothione levels in potassium antimonyl tartrate SbIII resistant *Leishmania tarentolae*. Mol.Biochem.Parasitol. 108: 131-135.

136. Haimeur, A., Guimond, C., Pilote, S., Mukhopadhyay, R., Rosen, B. P., Poulin, R., and Ouellette, M. 1999. Elevated levels of polyamines and trypanothione resulting from overexpression of the ornithine decarboxylase gene in arsenite- resistant *Leishmania*. Mol.Microbiol. 34: 726-735.

137. Gupta, N., Goyal, N., and Rastogi, A. K. 2001. In vitro cultivation and characterization of axenic amastigotes of *Leishmania*. Trends Parasitol 17: 150-153.

138. Bates, P. A. 1993. Axenic culture of *Leishmania* amastigotes. Parasitol Today 9: 143-146.

139. Zilberstein, D. and Shapira, M. 1994. The role of pH and temperature in the development of *Leishmania* parasites. Annu.Rev.Microbiol. 48: 449-470.

140. Haimeur, A. and Ouellette, M. 1998. Gene amplification in *Leishmania* tarentolae selected for resistance to sodium stibogluconate. Antimicrob.Agents Chemother. 42: 1689-1694.

141. Berdel, W. E., Fink, U., and Rastetter, J. 1987. Clinical phase I pilot study of the alkyl lysophospholipid derivative ET-18-OCH3. Lipids 22: 967-969.

142. Grunicke, H. H. and Uberall, F. 1992. Protein kinase C modulation. Semin.Cancer Biol. 3: 351-360.

143. Hilgard, P., Klenner, T., Stekar, J., and Unger, C. 1993. Alkylphosphocholines: a new class of membrane-active anticancer agents. Cancer Chemother.Pharmacol. 32: 90-95.

144. Beckers, T., Voegeli, R., and Hilgard, P. 1994. Molecular and cellular effects of hexadecylphosphocholine miltefosine in human myeloid leukaemic cell lines. Eur.J.Cancer 30A: 2143-2150.

145. Grunicke, H. H. 1998. Inhibition of phospholipase C and protein kinase C by alkylphosphocholines. Drugs Today 34: 3-14.

146. Lohmeyer, L. and Bittman, R. 1994. Antitumor ether-lupuds and alkyl-phosphocholines. Drugs Future 19v1021-1-37.

147. Schmidt-Ott, R., Klenner, T., Overath, P., and Aebischer, T. 1999. Topical treatment with hexadecylphosphocholine Miltex efficiently reduces parasite burden in experimental cutaneous leishmaniasis. Trans.R.Soc.Trop.Med.Hyg. 93: 85-90.

148. Perez-Victoria, J. M., Perez-Victoria, F. J., Parodi-Talice, A., Jimenez, I. A., Ravelo, A. G., Castanys, S., and Gamarro, F. 2001. Alkyl-lysophospholipid resistance in multidrug-resistant *Leishmania tropica* and chemosensitization by a novel P-glycoprotein-like transporter modulator. Antimicrob.Agents Chemother. 45: 2468-2474.

134   Zilberstein and Ephros

149. Lux, H., Heise, N., Klenner, T., Hart, D., and Opperdoes, F. R. 2000. Ether--lipid alkylphospholipid metabolism and the mechanism of action of ether--lipid analogues in *Leishmania*. Mol.Biochem.Parasitol. 111v 1-14.
150. Rybczynska, M., Liu, R., Lu, P., Sharom, F. J., Steinfels, E., Pietro, A. D., Spitaler, M., Grunicke, H., and Hofmann, J. 2001. MDR1 causes resistance to the antitumour drug miltefosine. Br.J.Cancer 84 : 1405-1411.
151. Beach, D. H., Goad, L. J., and Holz, G. G., Jr. 1988. Effects of antimycotic azoles on growth and sterol biosynthesis of *Leishmania* promastigotes. Mol.Biochem.Parasitol. 31v 149-162.
152. Mishra, M., Biswas, U. K., Jha, A. M., and Khan, A. B. 1994. Amphotericin versus sodium stibogluconate in first-line treatment of Indian kala-azar. Lancet 344: 1599-1600.
153. Mishra, M., Biswas, U. K., Jha, D. N., and Khan, A. B. 1992. Amphotericin versus pentamidine in antimony-unresponsive kala- azar. Lancet 340: 1256-1257.
154. Mbongo, N., Loiseau, P. M., Billion, M. A., and Robert-Gero, M. 1998. Mechanism of amphotericin B resistance in *Leishmania donovani* promastigotes. Antimicrob.Agents Chemother 42: 352-357.
155. Jha, T. K. 1983. Evaluation of diamidine compound pentamidine isethionate in the treatment resistant cases of kala-azar occurring in North Bihar, India. Trans.R.Soc.Trop.Med.Hyg. 77: 167-170.
156. Sundar, S. 2001. Drug resistance in Indian visceral leishmaniasis. Trop.Med.Int.Health 6: 849-854.
157. Basselin, M., Lawrence, F., and Robert-Gero, M. 1996. Pentamidine uptake in *Leishmania donovani* and *Leishmania* amazonensis promastigotes and axenic amastigotes. Biochemical.Journal. 315: 631-634.
158. Basselin, M., Coombs, G. H., and Barrett, M. P. 2000. Putrescine and spermidine transport in *Leishmania*. Mol.Biochem.Parasitol. 109: 37-46.
159. Berman, J. D., Gallalee, J. V., and Hansen, B. D. 1987. *Leishmania mexicana*: uptake of sodium stibogluconate Pentostam and pentamidine by parasite and macrophages. Exp.Parasitol. 64: 127-131.
160. Calonge, M., Johnson, R., Balana-Fouce, R., and Ordonez, D. 1996. Effects of cationic diamidines on polyamine content and uptake on *Leishmania infantum* in in vitro cultures. Biochem.Pharmacol. 52: 835-841.
161. Basselin, M., Badet-Denisot, M. A., Lawrence, F., and Robert-Gero, M. 1997. Effects of pentamidine on polyamine level and biosynthesis in wild- type, pentamidine-treated, and pentamidine-resistant *Leishmania*. Exp.Parasitol. 85: 274-282.
162. Basselin, M. and Robert Gero, M. 1998. Alterations in membrane fluidity, lipid metabolism, mitochondrial activity, and lipophosphoglycan expression in pentamidine-resistant *Leishmania*. Parasitol.Res. 84: 78-83.
163. Basselin, M., Lawrence, F., and Robert Gero, M. 1997. Altered transport properties of pentamidine-resistant *Leishmania donovani* and L. amazonensis promastigotes. Parasitol.Res. 83: 413-418.
164. Lopez-Jaramillo, P., Ruano, C., Rivera, J., Teran, E., Salazar-Irigoyen, R., Esplugues, J. V., and Moncada, S. 1998. Treatment of cutaneous leishmaniasis with nitric-oxide donor. Lancet 351: 1176-1177.
165. Davidson, R. N., Yardley, V., Croft, S. L., Konecny, P., and Benjamin, N. 2000. A topical nitric oxide-generating therapy for cutaneous leishmaniasis. Trans R Soc Trop Med Hyg. 94v319-322.
166. Arevalo, I., Ward, B., Miller, R., Meng, T. C., Najar, E., Alvarez, E., Matlashewski, G., and Llanos-Cuentas, A. 2001. Successful treatment of drug-resistant cutaneous leishmaniasis in humans by use of imiquimod, an immunomodulator. Clin.Infect.Dis. 33: 1847-1851.

APPENDIX
Overview of anti-leishmanial chemotherapy, (1-6):

SYNDROME	DRUG	DOSAGE (daily*, for normal adults)	COMMENTS
VL	Pentavalent antimony (sodium stibogluconate or meglumine antimoniate)	20mg Sb/kg for 28 days iv/im	Toxicity with higher daily doses or longer duration of therapy (e.g. PKDL); caveat: resistance e.g. Bihar. Monitor liver and renal function, pancreatic enzymes, QTc.
	Amphotericin B deoxycholate	0.5-1.0 mg/kg or every other day (total 15-20 mg/kg) iv	Regional variability, sometimes lower total doses; caveat: hypokalemia, thrombocytopenia, renal dysfunction, high fever.
	Amphotericin B fat emulsion	2 mg/kg iv every other day, 10 days	Generic formulation, relatively cheap. Caveat: standardization and quality control issues.
	Amphotericin B lipid complex	2 or 3 mg/kg iv, 5 days	5 mg/kg once is inadequate.
	Liposomal amphotericin B	iv 1 mg/kg, 5 days, or 5 or 7.5 mg/kg once	FDA approved regimens: 3 mg/kg/d on days 1-5, 14, 21; for immune compromised 4 mg/kg/d on days 1-5,10,17,24,31,38.
	Aminosidine	15-21 mg/kg, 21 days iv/im	Alone or in combination with other anti-leishmanial drugs.
	Pentamidine	4 mg/kg im/iv, 3 times/week, 15-30 doses	Toxic, possible vascular collapse, hypoglycemia/diabetes, renal failure, abscesses (if given im); suboptimal effectiveness.
	Miltefosine	50-100 mg/day po, by weight, for 28 days	Not for pregnant women, studies in progress. Non-Indian VL under study.
CL	Pentavalent antimony	20 mg/kg im/iv for 20 days	Shorter schedules by region/parasite, variable effectiveness.
	Amphotericin B deoxycholate	-	Infrequently used - only when specific situation warrants.
	Lipid Amphotericin formulations	-	May not be useful or warranted for CL.
	Pentamidine	3 mg/kg alternate	Focal studies only; less

		days for 4 doses or 2 mg/kg alternate days for 7 doses im/iv	toxicity due to shorter duration of therapy; may be good choice for *L. aethiopica.*
	Ketoconazole	200-800 mg, 28-56 days	For *L. mexicana*, *L. panamensis*, possibly *L. major.*
	Itraconazole	20 mg bid, 28 days	Sporadic successes.
	Allopurinol	200mg/kg qid, 28 days	Rare successes, new > old world.
	Dapsone	100-200 mg, 3-6 weeks	Old world leishmaniasis
	Intralesional antimony	Variable	Not always practical (number, location).
	Paromomycin/ and MBCL ointment	Twice daily applications, 10-20 days	Especially *L. major, L. mexicana.*
MCL	Pentavalent antimony	Dose and duration as per VL, or longer	Response variable.
	Amphotericin B deoxycholate	Dose as per VL, longer duration	Total dose of 20-40 mg/kg.
	Lipid Amphotericin formulations	-	May not be useful/warranted for MCL-inadequately studied to date.
	Pentamidine	2-4 mg/kg every other day, or 3 doses per wk for 15 doses at least	2nd line therapy.

*Total daily dose unless otherwise specified

THE IMMUNOLOGY OF VISCERAL LEISHMANIASIS: CURRENT STATUS

Paul M. Kaye

Department of Infectious and Tropical Diseases, London School of Hygiene and Tropical Medicine, Keppel Street, London, WC1E 7HT
United Kingdom

INTRODUCTION

The human impact of visceral leishmaniasis is significant. Some 500,000 new cases are reported annually, and epidemics may decimate local populations. A more complete understanding of the human immune response and its pathological consequences are needed for the design of vaccines and for improvements in therapy. This will require more thorough clinical investigation, and also a better appreciation of the usefulness and limitations of animal models. This review will concentrate on recent observations on patients with active VL. It will summarize recent progress in studies using a murine model that may have most bearing on human disease. Finally, it discusses future research directions aimed at elucidating the pathogenesis of VL.

THE CLINICAL IMMUNOLOGIST'S VIEW

Clinicians do not encounter most individuals exposed to *L. donovani* and *L. infantum*. Current epidemiological data suggests that greater than 95% of exposed immunocompetent individuals show no clinical signs of infection (1). There are no reports of the early kinetics of the immune response, nor of the precise mechanisms that control parasite multiplication and / or allow for the establishment of subclinical infection. Field observations in endemic regions indicate that recall proliferative and cytokine responses are primed (2). Genetics, age, malnutrition and concurrent infection are contributing factors in tipping the balance towards active disease (see later).

Active VL has a typical presentation. An incubation time with limited symptoms is followed by progressive splenomegaly, a lesser degree of hepatomegaly, intermittent or recurrent fever and cachexia. Lymphadenopathy is common in Africa and the Mediterranean, but not India (1,3-5). Pancytopenia is the norm, but there is variability in the relative impact on Hb, leucocytes and platelet count. Leucopenia is usually a result of loss of neutrophils, and there may be a relative lymphocytosis (1,4,6). Diminished *Leishmania*-specific recall responses, and increasing serum immunoglobulins are the norm. Response in PBMC cultures suggest that the immune response in human VL is of a mixed cytokine nature (4,7-9). Significantly, recall IFNγ responses are elevated by in vitro neutralization of IL-10. Conversely, addition of IL-12 enhnaces IFNγ prduction (10). An inhibitory role for IL-10 is supported by the observation that both IFNγ and IL-10 mRNA co-exist in fresh biopsy material from patients with active disease. Following treatment, IL-10 mRNA decreases, but the impact on IFNγ responses is variable (11,12). The cellular source of IL-10 in human VL has not been identified, and macrophages, T cells and B cells may all play a part. Specific subsets of T cells producing both IL-10 and IFNγ have been detected in recall responses of cured patients (13).

Serum cytokines have in general been in accord with the results of these studies. In Indian VL, Sundar and colleagues have made the most extensive study of human VL, their patients being in the major focus of the Bihar epidemic. These patients represent some of the most socio-economically underprivileged, with incomes of less than 1USD per day. IFNγ levels were comparable to controls, whereas IL-10 and IL-4 were elevated 3- and 13-fold, respectively in active cases. Post treatment, there were significant shifts in the IFNγ : IL-4 ratio, and IL-10 appeared to play a less dominant role than IL-4 (9) . In a study in northern Iran (5), IL-4 was not detected, but IL-13 provided a surrogate marker of Th2 activity. Surprisingly, there was a marked concordance in IL-13 and IFNγ production, though many patients with active disease had serum IL-10 in the absence of either of these two cytokines. Strikingly, of seven patients that continually relapsed following antimonial therapy, almost all had serum IL-10 in the absence of detectable serum IL-13 or IFNγ (5). Further prospective studies are required to determine whether the presence of IL-10 and absence of IFNγ / IL-13 in primary acute disease is predictive of relapse in Iran, as this did not appear to be the case in India (9). The expression of IL-10 by keratinocyes has recently been implicated in the evolution of PKDL (14).

THE GENETIC CONTROL OF HUMAN VISCERAL LEISHMANIASIS

The genetics underlying human resistance to *L. donovani* and *L. chagasi,* and the progression to VL has been the subject of intense analysis. The major gene controlling early rate of parasite growth in mice (Lsh/Nramp1/Slc11a1) has been thoroughly investigated (15-17). SLC11A1 / Slc11a1 is a proton / cation antiporter located in the phagosome membrane (17-19), and as a result of the central role of cation homeostasis in regulating cellular function, SLC11A1 / Slc11a1 regulates numerous macrophage functions, including regulation of TNFα production (reviewed in (17)). In man, the promoter region of SLC11A1 contains a Z-DNA repeat with four alleles (i-iv). Of these, only two occur at significant frequency: allele (ii) ≈ 0.25%; allele (iii) ≈ 0.75%. Allele (iii) drives higher expression of reporter constructs than allele (ii), providing a plausible functional basis for the observed association of allele (iii) with rheumatoid arthritis (20,21) and allele (ii) with tuberculosis (22,23). Surprisingly, a recent study identifies allele (iii) on a SLC11A1 haplotype associated with VL and PKDL in the Sudan (Blackwell, personal communication).

In northeastern Brazil, there is a high relative risk of VL in further siblings of infected sibling pairs and segregation analysis support single dominant or additive gene control (24). Although murine studies had indicated a role for MHC class II polymorphism in the regulation of cure rate (25), and polymorphisms in the promoter region of the TNFα gene have been suggested to contribute to progression of mucocutaneous leishmaniasis (26), a recent survey of 15 polymorphic loci across the HLA complex (spanning 1.3Mb from HLADQB1 to the TNFα microsatellite), failed to detect any association between HLA complex- encoded genes and susceptibility to VL in Brazil (Peacock, C.S. et.al. submitted). Results from other studies in Iran, India and the Mediterranean region also fail to support any strong association between susceptibility to VL and MHC class II or class III genes (27-29). Preliminary data from another study in northeastern Brazil, however, suggest that once infected with *L. chagasi*, an individual is more likely to develop VL if s/he has the TNF2 allele at position −308 of the TNFα gene. This allele is associated with higher levels of TNFα gene transcription and elevated resting serum TNFα concentrations. In contrast, individuals with the TNF1 allele are more likely to develop asymptomatic infection (Wilson, personal communication). Collectively, these genetic data suggest that unlike the situation in the mouse model of VL (30,31), there may be some therapeutic benefit in interventions targeted at TNFα (32).

THE IMMUNOPATHOLOGY OF HUMAN VL

Detailed histopathological data is scant (4,6). Enlargement of the spleen is one of the defining features of VL. There is white pulp atrophy, loss of T cells from the periarteriolar region and a widespread infiltration of plasma cells and infected macrophages in both the white pulp and red pulp. Lymph nodes, if involved, show follicular destruction, lack of germinal centres and replacement of paracortical T cells with parasitised macrophages. Epitheliod granulomas, characteristic of cutaneous leishmaniasis (33), are not observed. Both the spleen and lymph nodes may show evidence of immune complex deposition. The liver is also enlarged, with Kupffer cell hypertrophy and hyperplasia. Infected Kupffer cells, as well as macrophages in the portal tracts are readily detected, and there may be evidence of fibrosis. The role of the granulomatous response in protection from VL is inferred from the identification of small granulomas in the liver of subclinical cases (34,35), and from studies of the granulomatous pathology of other leishmanial diseases (33). However, it should be noted that acute VL, unlike in the mouse model, rarely presents with granulomas in the liver (1,4,6). There have been no recent immunohistological studies of VL, in which cellular phenotype has been characterized using modern immunological markers. Given the clear importance of tissue microenvironment in governing anti-leishmanial responses (see below), more data on the nature of the local immune response in active disease is clearly needed.

IMMUNOSUPPRESSION IN HUMAN VL

As mentioned above, there is a significant literature suggestive of an immunosuppressive role for IL-10. However, other contributing mechanisms are possible. Cillari et al noted a dramatic loss of CD45RO+ CD4+ T cells in PBMC from patients with VL caused by L. infantum in Sicily. The frequency of these cells returned to normal range within 6 weeks of successful therapy (7). This indicates selective loss of effector / memory CD4+ T cells, either by activation induced cell death or by other apoptotic pathways. Although apoptosis can be driven via CD95 / CD95L interactions, in many infectious diseases including murine infection with L. donovani (36) and L. major (37), a role for TNFα has been suggested. The elevated levels of TNFα seen in VL patients support a model of T cell loss due to elevated TNFα mediated apoptosis. In addition, these results beg the question of whether individuals with active VL have the capacity to replenish their effector cell pool, or whether this function only returns with the onset of effective chemotherapy. Intriguingly, a recent study has shown that the restoration of lymphocyte responsiveness, as a result of drug treatment, may predispose to PKDL, with effector function then targeted against persistent parasites in the skin (38). Surprisingly, there have been no reports to date of the role of TGFβ in human VL, though with the high level of attention focused on this cytokine this data is likely to be forthcoming. Some evidence suggests however that regulatory subsets of T cells can be expanded from patients following active VL and that these are able to make both IFNγ and IL-10 (13).

Epidemiological studies indicate that the risk of VL is linked to both age and nutritional status (39). However, whilst these are likely to impart some degree of immunological deviation or immunosuppression, there have been no formal studies of age-dependent or nutrition-dependent changes in immune responses to visceralising species of *Leishmania*. More severe immunosuppression, associated with HIV infection, is however a clear risk factor (40), with increasing significance as areas of HIV transmission extend to overlap that of VL.

THE MURINE MODEL OF VL: LESSONS OR DISTRACTIONS?

Mouse models have a major influence in shaping the interpretation of the response in human VL. Whilst there are undoubtedly some similarities and valid extrapolations to be made (some of which are highlighted in the following sections), it is important to keep the information from this model in context.

Much of our understanding of the immune response to the visceralising species of *Leishmania* comes from studies in which amastigotes are administered intravenously into the naïve, well-nourished rodent host. There have to date been few experimental infections initiated by the natural dermal route or even using metacyclic promastigotes. In an elegant study combining low dose infection with selective nutrient deficiency, Melby and colleagues recently demonstrated that the lymph nodes draining the infection site play an important role in preventing dissemination of *L. donovani* (41). In mice malnourished to an extent similar to that seen in disease endemic populations, this barrier function was reduced and parasite dissemination occurred more rapidly into systemic sites. Wilson and colleagues have recently demonstrated that heavily attenuated *L. chagasi* promastigotes (derived by continuous passage or genetic attenuation) failed to induce immunity to systemic challenge (42). In contrast, immunity was achieved by high doses of virulent promastigotes, though these failed to spread systemically. Together, these two studies suggest that the containment and local growth of virulent parasites can induce protective immunity expressed in the viscera. This may be the natural route to generate resistance in man, and should local parasites not be sterilized, may also provide a pool of parasites able to form the target for PKDL In contrast, if amastigotes (or amastigote-infected macrophages or dendritic cells) spill directly to systemic sites, this may seed progressive disease. Whether the outcome of infection truly reflects competition between locally primed and systemically primed immune responses, or merely inefficient operation of locally primed responses in a visceral microenvironment is open to further study. Likewise, these experiments suggest that investigation of lymph barrier function in endemic areas of VL may be profitable. Further comparative studies of murine models of VL, using some of the approaches recently applied to murine cutaneous disease (see Chapter 11 by Farrell) are also likely to be of importance in defining local vs. systemic responses.

THE OUTCOME OF HEPATIC INFECTION

The outcome of infection in the murine liver has been well documented and the subject of major recent reviews (1,43). The earliest response noted is the initiation of chemokine production as a direct result of KC infection (44). MIP-1α and MCP-1 mRNAs are transiently induced (peak 5h) in wild type (WT) and SCID mice, whereas mRNA for γIP-10 follows similar early kinetics but is sustained in the presence of T cells. This study indicated that transient expression of MIP1α, MCP-1 and γIP-10, in the absence of T cells, could not induce local inflammation and early granuloma formation. However, initial granuloma diameter is reduced in MIP1α-/- and CCR2-/- mice, but not in CCR5-/- mice (45). More long-term studies on the regulation of granuloma formation using chemokine and chemokine receptor KO mice are complicated by changes in lymphocyte function (45). It is not known whether liver resident T cells are actively involved in the regulation of hepatic disease, one interpretation of the studies of Cotterell et al. (44), and there has been surprisingly little information on the role of non-classical T cells (46). The liver has its own complement of antigen presenting cells, and the specialization of hepatic DC populations (47) and of hepatic endothelium (48) make this an exciting area for future research. Indirect evidence suggests that priming may occur elsewhere, notably in the spleen and paradoxically that such T cells are effective in the hepatic but not the splenic microenvironment (49).

The development of the hepatic granuloma can be staged histologically (43). This proceeds with the initial fusion of infected Kupffer cells, the recruitment of a predominantly mononuclear infiltrate and the eventual "maturation" of the infiltrate into an organised structure. CD4+ T cells predominate early in this process, with CD8+ T cells appearing later. B cells are scarce, as are neutrophils. Nevertheless, Gr-1+ neutrophils have recently been shown to be an essential component of the host response (50,51). It is unclear whether this relates to their effector function, or to regulation of downstream events in inflammation. Studies with phox47-/- and NOS2-/- mice indicate that both oxygen radicals and nitric oxide are necessary to contain parasite numbers in this organ (52). Although Gr-1 is predominantly expressed on granulocytes, a recently identified subset of DC, similar to human plasmacytoid DC, also expresses Gr-1 (53). As these cells have the capacity to produce type I IFN, their role in regulating early events in visceral infection needs to be clarified. The requirement for efficient granuloma formation has been examined using various gene targeted mice and by the administration of neutralizing mAbs (43). The importance of IL-10 in dampening anti-parasite responses has recently been examined using IL-10-/- mice (54). Early hepatic parasite burden was markedly reduced at 7 days post infection and had cleared by day 21, the peak of the infection in WT mice. Increased resistance was accompanied by elevated levels of INFγ and NO. Surprisingly, the authors report little adverse pathology as a result of this heightened Th1-like response. IL-12 has a key role in regulating parasite burden in IL-10-/- mice, acting through both IFNγ-dependent and IFNγ-independent pathways. These finding are entirely consistent with our concepts of macrophage activation for leishmanicidal activity (55). The role of IL-4 in murine visceral infection is still enigmatic. IL-4-/- mice have marginally elevated parasite loads early in infection (56,57) and a delayed long-term clearance. Granuloma formation is depressed, though this is more marked in IL-4Rα-/- mice, where IL-13 signaling is also impaired (Stager et.al. unpublished). Recent studies also indicate that the presence of IL-4 is essential for the induction of vaccine induced immunity, this reflecting the role of IL-4 in regulating CD8+ T cell function (58).

In vivo blockade of CTLA-4 (CD152) has been shown to enhance clearance of *L. donovani* from the liver, whether given immediately after infection or as a late therapeutic agent (59) (and unpublished). Whilst it is tempting to speculate that this reflects the role of CTLA-4 in the regulation of TGFβ production (reviewed in 60), studies to discriminate CTLA-4 regulation of TGFβ from the known role of CTLA-4 as a direct negative regulator of T cell activation (61) have not been performed. Nevertheless, TGFβ has been implicated the regulation of hepatic parasite burden and in controlling local T cell function in *L. chagasi* infection (62).

THE INDUCTION OF ANTI-LEISHMANIAL IMMUNITY IN THE SPLEEN

The immune response in the mouse liver may well represent that seen in subclinical VL, limiting its usefulness in understanding progressive human disease. Recent studies have begun to appreciate that the mouse spleen provides a more accurate representation of these events. Histological studies identify the marginal zone macrophages and the marginal metallophilic macrophages as initial targets of infection (63). These are likely to produce chemokines though their repertoire may be more limited than that of Kupffer cells (64). In situ analysis also identified dendritic cells as the immediate (5-24h) source of IL-12 ((63) and Dianda et.al., submitted). Most microbial pathogens stimulate IL-12 production in DC by interacting with Toll-like Receptors (65), but this direct response is not observed with *L. donovani*. Amastigotes fail to trigger IL-12 from DC in vitro, or in situ in SCID, RAG-/- or CD3εTg25 mice. Indeed, as with L. mexicana infection (66), infection of DC by *L. donovani* is not followed by increased maturation of DC (Dianda et.al., submitted). In contrast, there is a strict requirement for CD4+ T cells, which induce IL-12 via a mechanism involving both TCR ligation and an additional costimulus, distinct from CD40-CD40L interactions (submitted). An analysis of the numbers of DC that are activated to IL-12 production and of the precursor frequency of *Leishmania* specific T cells, leads to the conclusion that cross-reactive T cells may play a role in early stages of immunoregulation following this infection. Chemokines also play an important part in the regulation of DC-T cell interactions, allowing their migration and mutual interaction within the T cell area of the white pulp. Our initial studies indicate that migration of DC into this area during L. donovani infection is facilitated by an increase in the frequency of DC that migrate in response to CCRL21 (= secondary lymphoid chemokine, SLC), a chemokine produced by the stromal cells of the PALS and by the endothelium of the central arteriole (67). There has been little characterization of the antigen specificity of the response to L. donovani, though unlike *L. major* infection, early expansion of a LACK-reactive CD4+ T cell population does not occur (68). Perhaps as a consequence, *L. donovani* infection does not induce the Th subset polarization seen following L. major infection. Thus, sensitive ELISPOT analysis indicated that both IL-4 and IFNγ are produced following infection (59). Surprisingly, IL-12 is required for efficient priming of both these Th subsets (69).

THE PATHOLOGY OF MURINE VISCERAL LEISHMANIASIS

Although T cell priming likely occurs within the specialized microarchitecture of the splenic white pulp, the long-term outcome of infection in the spleen is distinct from that seen in the liver. Thus, after a relative quiescent period when parasite growth is hard to detect (until day 28pi) there is a dramatic increase in parasite load, eventually reaching a plateau which may persist for the life of the animal (70,71). A numbers of pathological changes develop, which bear remarkable similarity to those seen in human VL (71). Additionally, there is marked elevation of local hematopoietic activity (64,72). In relation to Th subset balance and the local cytokine milieu it has been surprisingly difficult to identify any distinguishing features of the late response in the spleen and that in the liver during cure. We have recently looked for other ways to explain the diverse outcomes in these tissues. Various studies have shown that apoptosis is common in T cells isolated from mice infected with *L. donovani* (36,73,74), and the relative increase in apoptosis duriing infection is most marked in the spleen (36). This may reduce effector / memory cell numbers significantly as infection progresses. Second, we noted a number of key structural changes in this organ. Marginal zone macrophages undergo selective destruction during chronic infection, a process mediated via a TNFα (Engwerda, et.al., submitted). This removes a critical route for efficient lymphocyte migration into the PALS. We also noted that the key T zone stromal elements are also lost during chronic disease, leading to reduction of SLC expression in the PALS (Ato et.al., submitted). Together, these changes result in an almost complete failure of T cell to traffic into the infected spleen. Importantly, mice are protected against many of these changes by chemotherapy, providing a novel explanation for the recovery of effector cells post therapy in man.

OTHER MODELS OF IMMUNOSUPPRESSION IN THE SPLEEN

CTLA-4 blockade, unlike many other interventions, also has a significant impact on splenic parasite load (59). DosReis and colleagues have provided strong supportive data that within this microenvironment, CTLA-4 function by regulating TGFβ production. IFNγ responses (and to a lesser extent IL-4 responses) can be restored in splenic T cells taken from chronically infected mice by addition of anti-CTLA-4 mAb in vitro (75). This effect was mimicked by anti-TGFβ antibody, and conversely cross-linking of CTLA-4 led to enhanced TGFβ production in these cultures. Although TGFβ inhibits T cell cytokine production, this may not be the only axis by which TGFβ exerts its effects. TGFβ induces ornithine decarboxylase in macrophages, and the increased level of polyamine synthesis may promote growth of *Leishmania* as it does for intracellular Trypanosoma cruzi (60). The cellular source of TGFβ has not been formally identified but in vivo depletion of CD25+ T cells enhances parasite clearance in a similar manner to CTLA-4 blockade (Dianda et.al., in preparation), suggesting a possible role for CD4+CD25+CTLA-4+ T regulatory cells. IL-10-/- mice also show reduced splenic parasite burden and elevated T cell responses in vitro (54). However therapeutic studies with neutralizing anti-IL-10 mAbs have been equivocal. Surprisingly, as in man, there has been no formal identification of the cells responsible for IL-10 production.

THE HAMSTER MODEL OF VISCERAL LEISHMANIASIS

Hamsters are uniquely susceptible to *L. donovani* infection, and are the only small animal model to acquire disease similar to that seen in man. Thus, whilst both hamster and mouse develop overt Th1 biased immune responses, and show local pathology akin to that seen in the human disease, only the hamster progresses to a fatal outcome with associated cachexia. It now emerges that the principal defect in the hamster immune response lies at the effector phase of the immune response, namely in deficient induction of NOS2. It is possible that the hamster NOS promoter has similar structural characteristics to that of human NOS2, which make it less responsive to IFNγ (76). This study is important in that it highlights both the importance of choice of model as interventions become nearer to reality, and in that it indicates that our preoccupation with early instructional events in immune regulation may underestimate the necessity to have appropriately intact effector pathways.

CONCLUDING REMARKS

There is no doubt that research using murine models of disease will continue to play a role in formulating our concepts of immunity and pathology in VL. But by placing greater emphasis in the future on dissecting the immunology of human disease, we will also gain a greater understanding of the true usefulness of the model (notably in assessing otherwise intractable questions relating to systemic immunity). This will be come more important as potential vaccine candidates and novel immunotherapeutics are developed and require pre-clinical evaluation.

ACKNOWLEDGEMENTS

Work in the author's laboratory is supported by grants from the Wellcome Trust and the British Medical Research Council. Thanks go to Jennie Blackwell, Mary Wilson, George DosReis and Jay Farrell for sharing unpublished data, and to all my colleagues who have been involved in the work reported here.

REFERENCES
1.Baker, R., Chiodini, P., and Kaye, P. M. 1999. Leishmaniasis. In Geraint James, D. and Zumla, A., eds., The Granulomatous Disorders, p. 212 - 234. Cambridge University Press, Cambridge.
2.Sacks, D. L., Lal, S. L., Shrivastava, S. N., Blackwell, J., and Neva, F. A. 1987. An analysis of T cell responsiveness in Indian kala-azar. J Immunol 138:908-13.
3.Thakur, C. P. Epidemiological, clinical and therapeutic features of Bihar kala-azar (including post kala-azar dermal leishmaniasis).
4.Zijlstra, E. E. and el-Hassan, A. M. 2001. Leishmaniasis in Sudan. Visceral leishmaniasis. Trans R Soc Trop Med Hyg 95 Suppl 1:S27-58.
5.Babaloo, Z., Kaye, P. M., and Eslami, M. B. 2001. Interleukin-13 in Iranian patients with visceral leishmaniasis: relationship to other Th2 and Th1 cytokines. Trans R Soc Trop Med Hyg 95:85-8.
6.Bryceson, A. D. M. 1996. Manson's Tropical Diseases, 20 edn. W.B. Saunders Ltd., London.
7.Cillari, E., Vitale, G., Arcoleo, F., D'Agostino, P., Mocciaro, C., Gambino, G., Malta, R., Stassi, G., Giordano, C., Milano, S., and et al. 1995. In vivo and in vitro cytokine profiles and mononuclear cell subsets in Sicilian patients with active visceral leishmaniasis. Cytokine 7:740-5.
8.Reed, S. G. and Scott, P. 1993. T-cell and cytokine responses in leishmaniasis. Curr Opin Immunol 5:524-31.

9.Sundar, S., Reed, S. G., Sharma, S., Mehrotra, A., and Murray, H. W. 1997. Circulating T helper 1 (Th1) cell- and Th2 cell-associated cytokines in Indian patients with visceral leishmaniasis. Am J Trop Med Hyg 56:522-5.

10.Ghalib, H. W., Whittle, J. A., Kubin, M., Hashim, F. A., el-Hassan, A. M., Grabstein, K. H., Trinchieri, G., and Reed, S. G. 1995. IL-12 enhances Th1-type responses in human Leishmania donovani infections. J Immunol 154:4623-9.

11.Karp, C. L., el-Safi, S. H., Wynn, T. A., Satti, M. M., Kordofani, A. M., Hashim, F. A., Hag-Ali, M., Neva, F. A., Nutman, T. B., and Sacks, D. L. 1993. In vivo cytokine profiles in patients with kala-azar. Marked elevation of both interleukin-10 and interferon-gamma [see comments]. J Clin Invest 91:1644-8.

12.Kenney, R. T., Sacks, D. L., Gam, A. A., Murray, H. W., and Sundar, S. 1998. Splenic cytokine responses in Indian kala-azar before and after treatment. J Infect Dis 177:815-8.

13.Kemp, K., Kemp, M., Kharazmi, A., Ismail, A., Kurtzhals, J. A., Hviid, L., and Theander, T. G. 1999. Leishmania-specific T cells expressing interferon-gamma (IFN-gamma) and IL-10 upon activation are expanded in individuals cured of visceral leishmaniasis. Clin Exp Immunol 116:500-4.

14.Gasim, S., Elhassan, A. M., Khalil, E. A., Ismail, A., Kadaru, A. M., Kharazmi, A., and Theander, T. G. 1998. High levels of plasma IL-10 and expression of IL-10 by keratinocytes during visceral leishmaniasis predict subsequent development of post- kala-azar dermal leishmaniasis. Clin Exp Immunol 111:64-9.

15.Bradley, D. J., Taylor, B. A., Blackwell, J., Evans, E. P., and Freeman, J. 1979. Regulation of Leishmania populations within the host. III. Mapping of the locus controlling susceptibility to visceral leishmaniasis in the mouse. Clin Exp Immunol 37:7-14.

16.Vidal, S., Tremblay, M. L., Govoni, G., Gauthier, S., Sebastiani, G., Malo, D., Skamene, E., Olivier, M., Jothy, S., and Gros, P. 1995. The Ity/Lsh/Bcg locus: natural resistance to infection with intracellular parasites is abrogated by disruption of the Nramp1 gene. J Exp Med 182:655-66.

17.Blackwell, J., Goswami, T., Evans, C. A. W., Sibthorpe, D., Papo, N., White, J. K., Searle, S., Miller, E. N., Peacock, C. S., Mohammed, H., and Ibrahim, M. 2001. SLC11A1 (formerly NRAMP1) and disease resistance. Cell Microbiol 3:in press.

18.Gruenheid, S., Pinner, E., Desjardins, M., and Gros, P. 1997. Natural resistance to infection with intracellular pathogens: the Nramp1 protein is recruited to the membrane of the phagosome. J Exp Med 185:717-30.

19.Vidal, S. M., Pinner, E., Lepage, P., Gauthier, S., and Gros, P. 1996. Natural resistance to intracellular infections: Nramp1 encodes a membrane phosphoglycoprotein absent in macrophages from susceptible (Nramp1 D169) mouse strains. J Immunol 157:3559-68.

20.Shaw, M. A., Clayton, D., Atkinson, S. E., Williams, H., Miller, N., Sibthorpe, D., and Blackwell, J. M. 1996. Linkage of rheumatoid arthritis to the candidate gene NRAMP1 on 2q35. J Med Genet 33:672-7.

21.Sanjeevi, C. B., Miller, E. N., Dabadghao, P., Rumba, I., Shtauvere, A., Denisova, A., Clayton, D., and Blackwell, J. M. 2000. Polymorphism at NRAMP1 and D2S1471 loci associated with juvenile rheumatoid arthritis. Arthritis Rheum 43:1397-404.

22.Gao, P. S., Fujishima, S., Mao, X. Q., Remus, N., Kanda, M., Enomoto, T., Dake, Y., Bottini, N., Tabuchi, M., Hasegawa, N., Yamaguchi, K., Tiemessen, C., Hopkin, J. M., Shirakawa, T., and Kishi, F. 2000. Genetic variants of NRAMP1 and active tuberculosis in Japanese populations. International Tuberculosis Genetics Team. Clin Genet 58:74-6.

23.Bellamy, R., Ruwende, C., Corrah, T., McAdam, K. P., Whittle, H. C., and Hill, A. V. 1998. Variations in the NRAMP1 gene and susceptibility to tuberculosis in West Africans. N Engl J Med 338:640-4.

24.Peacock, C. S., Collins, A., Shaw, M. A., Silveira, F., Costa, J., Coste, C. H., Nascimento, M. D., Siddiqui, R., Shaw, J. J., and Blackwell, J. M. 2001. Genetic epidemiology of visceral leishmaniasis in northeastern Brazil. Genet Epidemiol 20:383-96.

25.Blackwell, J. M. Leishmania donovani infection in heterozygous and recombinant H-2 haplotype mice.

26.Cabrera, M., Shaw, M. A., Sharples, C., Williams, H., Castes, M., Convit, J., and Blackwell, J. M. 1995. Polymorphism in tumor necrosis factor genes associated with mucocutaneous leishmaniasis. J Exp Med 182:1259-64.

27.Meddeb-Garnaoui, A., Gritli, S., Garbouj, S., Ben Fadhel, M., El Kares, R., Mansour, L., Kaabi, B., Chouchane, L., Ben Salah, A., and Dellagi, K. 2001. Association analysis of HLA-class II and class III gene polymorphisms in the susceptibility to mediterranean visceral leishmaniasis. Hum Immunol 62:509-17.

28.Faghiri, Z., Tabei, S. Z., and Taheri, F. 1995. Study of the association of HLA class I antigens with kala-azar. Hum Hered 45:258-61.

29.Singh, N., Sundar, S., Williams, F., Curran, M. D., Rastogi, A., Agrawal, S., and Middleton, D. 1997. Molecular typing of HLA class I and class II antigens in Indian kala-azar patients. Trop Med Int Health 2:468-71.

30.Tumang, M. C., Keogh, C., Moldawer, L. L., Helfgott, D. C., Teitelbaum, R., Hariprashad, J., and Murray, H. W. 1994. Role and effect of TNF-alpha in experimental visceral leishmaniasis. J Immunol 153:768-75.

31.Murray, H. W., Jungbluth, A., Ritter, E., Montelibano, C., and Marino, M. W. 2000. Visceral leishmaniasis in mice devoid of tumor necrosis factor and response to treatment. Infect Immun 68:6289-93.

32.Gardnerova, M., Blanque, R., and Gardner, C. R. 2000. The use of TNF family ligands and receptors and agents which modify their interaction as therapeutic agents. Curr Drug Targets 1:327-64.

33.Ridley, D. S. A histological classification of cutaneous leishmaniasis and its geographical expression.

34.Pampiglione, S., Manson-Bahr, P. E., Giungi, F., Giunti, G., Parenti, A., and Canestri Trotti, G. Studies on Mediterranean leishmaniasis. 2. Asymptomatic cases of visceral leishmaniasis.

35.Daneshbod, K. 1972. Visceral leishmaniasis (kala-azar) in Iran: a pathologic and electron microscopic study. Am J Clin Pathol 57:156-66.

36.Alexander, C. E., Kaye, P. M., and Engwerda, C. R. 2001. CD95 is required for the early control of parasite burden in the liver of Leishmania donovani-infected mice. Eur J Immunol 31:1199-210.

37.Kanaly, S. T., Nashleanas, M., Hondowicz, B., and Scott, P. 1999. TNF receptor p55 is required for elimination of inflammatory cells following control of intracellular pathogens. J Immunol 163:3883-9.

38.Gasim, S., Elhassan, A. M., Kharazmi, A., Khalil, E. A., Ismail, A., and Theander, T. G. 2000. The development of post-kala-azar dermal leishmaniasis (PKDL) is associated with acquisition of Leishmania reactivity by peripheral blood mononuclear cells (PBMC). Clin Exp Immunol 119:523-9.

39.Dye, C. and Williams, B. G. 1993. Malnutrition, age and the risk of parasitic disease: visceral leishmaniasis revisited. Proc R Soc Lond B Biol Sci 254:33-9.

40.Pintado, V., Martin-Rabadan, P., Rivera, M. L., Moreno, S., and Bouza, E. 2001. Visceral leishmaniasis in human immunodeficiency virus (HIV)-infected and non-HIV-infected patients. A comparative study. Medicine (Baltimore) 80:54-73.

41.Anstead, G. M., Chandrasekar, B., Zhao, W., Yang, J., Perez, L. E., and Melby, P. C. 2001. Malnutrition alters the innate immune response and increases early visceralization following Leishmania donovani infection. Infect Immun 69:4709-18.

42.Streit, J. A., Recker, T. J., Filho, F. G., Beverley, S. M., and Wilson, M. E. 2001. Protective immunity against the protozoan Leishmania chagasi is induced by subclinical cutaneous infection with virulent but not avirulent organisms. J Immunol 166:1921-9.

43.Murray, H. W. 1999. Granulomatous inflammation: Host antimicrobial defense in the tissues in visceral leishmaniasis. In Gallin, J. I. and Snyderman, R., eds., Inflammation: basic principles and clinical correlates., p. 977-994. Lippincott Williams & Wilkins, Philadelphia.

44.Cotterell, S. E., Engwerda, C. R., and Kaye, P. M. 1999. Leishmania donovani infection initiates T cell-independent chemokine responses, which are subsequently amplified in a T cell-dependent manner. Eur J Immunol 29:203-14.

45.Sato, N., Kuziel, W. A., Melby, P. C., Reddick, R. L., Kostecki, V., Zhao, W., Maeda, N., Ahuja, S. K., and Ahuja, S. S. 1999. Defects in the generation of IFN-gamma are overcome to control infection with Leishmania donovani in CC chemokine receptor (CCR) 5-, macrophage inflammatory protein-1 alpha-, or CCR2-deficient mice. J Immunol 163:5519-25.

46.Crispe, I. N. and Mehal, W. Z. 1996. Strange brew: T cells in the liver. Immunol Today 17:522-5.

47.Trobonjaca, Z., Leithauser, F., Moller, P., Schirmbeck, R., and Reimann, J. 2001. Activating immunity in the liver. I. Liver dendritic cells (but not hepatocytes) are potent activators of IFN-gamma release by liver NKT cells. J Immunol 167:1413-22.

48.Knolle, P. A. and Limmer, A. 2001. Neighborhood politics: the immunoregulatory function of organ-resident liver endothelial cells. Trends Immunol 22:432-7.

49.Engwerda, C. R. and Kaye, P. M. 2000. Organ-specific immune responses associated with infectious disease. Immunol Today 21:73-8.

50.Smelt, S. C., Cotterell, S. E., Engwerda, C. R., and Kaye, P. M. 2000. B cell-deficient mice are highly resistant to Leishmania donovani infection, but develop neutrophil-mediated tissue pathology. J Immunol 164:3681-8.

51.Rousseau, D., Demartino, S., Ferrua, B., Francois Michiels, J., Anjuere, F., Fragaki, K., Le Fichoux, Y., and Kubar, J. In vivo involvement of polymorphonuclear neutrophils in Leishmania infantum infection.

52.Murray, H. W. and Nathan, C. F. 1999. Macrophage microbicidal mechanisms in vivo: reactive nitrogen versus oxygen intermediates in the killing of intracellular visceral Leishmania donovani. J Exp Med 189:741-6.

53.Nakano, H., Yanagita, M., and Gunn, M. D. 2001. Cd11c(+)b220(+)gr-1(+) cells in mouse lymph nodes and spleen display characteristics of plasmacytoid dendritic cells. J Exp Med 194:1171-8.

54.Murphy, M. L., Wille, U., Villegas, E. N., Hunter, C. A., and Farrell, J. P. 2001. IL-10 mediates susceptibility to Leishmania donovani infection. Eur J Immunol 31:2848-56.

55.Alexander, J., Satoskar, A. R., and Russell, D. G. 1999. Leishmania species: models of intracellular parasitism. J Cell Sci 112 Pt 18:2993-3002.

56.Alexander, J., Carter, K. C., Al-Fasi, N., Satoskar, A., and Brombacher, F. 2000. Endogenous IL-4 is necessary for effective drug therapy against visceral leishmaniasis. Eur J Immunol 30:2935-43.

57.Satoskar, A., Bluethmann, H., and Alexander, J. 1995. Disruption of the murine interleukin-4 gene inhibits disease progression during Leishmania mexicana infection but does not increase control of Leishmania donovani infection. Infect Immun 63:4894-9.

58.Stager, S., Smith, D. F., and Kaye, P. M. 2000. Immunization with a recombinant stage-regulated surface protein from Leishmania donovani induces protection against visceral leishmaniasis. J Immunol 165:7064-71.

59.Murphy, M. L., Cotterell, S. E., Gorak, P. M., Engwerda, C. R., and Kaye, P. M. 1998. Blockade of CTLA-4 enhances host resistance to the intracellular pathogen, Leishmania donovani. J Immunol 161:4153-60.

60.Gomes, N. A. and DosReis, G. A. 2001. The dual role of CTLA-4 in Leishmania infection. Trends in Parasitology 17:487 - 491.

61.Sullivan, T. J., Letterio, J. J., van Elsas, A., Mamura, M., van Amelsfort, J., Sharpe, S., Metzler, B., Chambers, C. A., and Allison, J. P. 2001. Lack of a role for transforming growth factor-beta in cytotoxic T lymphocyte antigen-4-mediated inhibition of T cell activation. Proc Natl Acad Sci U S A 98:2587-92.

62.Wilson, M. E., Young, B. M., Davidson, B. L., Mente, K. A., and McGowan, S. E. 1998. The importance of TGF-beta in murine visceral leishmaniasis. J Immunol 161:6148-55.

63.Gorak, P. M., Engwerda, C. R., and Kaye, P. M. 1998. Dendritic cells, but not macrophages, produce IL-12 immediately following Leishmania donovani infection. Eur J Immunol 28:687-95.

64.Cotterell, S. E., Engwerda, C. R., and Kaye, P. M. 2000. Enhanced hematopoietic activity accompanies parasite expansion in the spleen and bone marrow of mice infected with Leishmania donovani. Infect Immun 68:1840-8.

65.Reis e Sousa, C., Sher, A., and Kaye, P. 1999. The role of dendritic cells in the induction and regulation of immunity to microbial infection. Curr Opin Immunol 11:392-9.

66.Bennett, C. L., Misslitz, A., Colledge, L., Aebischer, T., and Blackburn, C. C. 2001. Silent infection of bone marrow-derived dendritic cells by Leishmania mexicana amastigotes. Eur J Immunol 31:876-83.

67.Luther, S. A., Tang, H. L., Hyman, P. L., Farr, A. G., and Cyster, J. G. 2000. Coexpression of the chemokines ELC and SLC by T zone stromal cells and deletion of the ELC gene in the plt/plt mouse. Proc Natl Acad Sci U S A 97:12694-9.

68.Launois, P., Ohteki, T., Swihart, K., MacDonald, H. R., and Louis, J. A. 1995. In susceptible mice, Leishmania major induce very rapid interleukin-4 production by CD4+ T cells which are NK1.1. Eur J Immunol 25:3298-307.

69.Engwerda, C. R., Murphy, M. L., Cotterell, S. E., Smelt, S. C., and Kaye, P. M. 1998. Neutralization of IL-12 demonstrates the existence of discrete organ- specific phases in the control of Leishmania donovani. Eur J Immunol 28:669-80.

70.Wilson, M. E., Sandor, M., Blum, A. M., Young, B. M., Metwali, A., Elliott, D., Lynch, R. G., and Weinstock, J. V. 1996. Local suppression of IFN-gamma in hepatic granulomas correlates with tissue-specific replication of Leishmania chagasi. J Immunol 156:2231-9.

71.Smelt, S. C., Engwerda, C. R., McCrossen, M., and Kaye, P. M. 1997. Destruction of follicular dendritic cells during chronic visceral leishmaniasis. J Immunol 158:3813-21.

72.Cotterell, S. E., Engwerda, C. R., and Kaye, P. M. 2000. Leishmania donovani infection of bone marrow stromal macrophages selectively enhances myelopoiesis, by a mechanism involving GM-CSF and TNF-alpha. Blood 95:1642-51.

73.Das, G., Vohra, H., Saha, B., Agrewala, J. N., and Mishra, G. C. Leishmania donovani infection of a susceptible host results in apoptosis of Th1-like cells: rescue of anti-leishmanial CMI by providing Th1-specific bystander costimulation.

74.Das, G., Vohra, H., Rao, K., Saha, B., and Mishra, G. C. 1999. Leishmania donovani infection of a susceptible host results in CD4+ T- cell apoptosis and decreased Th1 cytokine production. Scand J Immunol 49:307-10.

75.Gomes, N. A., Barreto-de-Souza, V., Wilson, M. E., and DosReis, G. A. 1998. Unresponsive CD4+ T lymphocytes from Leishmania chagasi-infected mice increase cytokine production and mediate parasite killing after blockade of B7-1/CTLA-4 molecular pathway. J Infect Dis 178:1847-51.

76.Melby, P. C., Chandrasekar, B., Zhao, W., and Coe, J. E. 2001. The hamster as a model of human visceral leishmaniasis: progressive disease and impaired generation of nitric oxide in the face of a prominent Th1-like cytokine response. J Immunol 166:1912-20.

THE IMMUNOLOGY OF CUTANEOUS LEISHMANIASIS: EXPERIMENTAL INFECTIONS AND HUMAN DISEASE

Jay P. Farrell

Department of Pathobiology, School of Veterinary Medicine, University of Pennsylvania, 3800 Spruce Street, Philadelphia, PA 19104

INTRODUCTION

Cutaneous leishmaniasis (CL) is a term used to describe a group of diseases caused by multiple species of parasites within the genus *Leishmania*. Most cases of cutaneous leishmaniasis will spontaneously heal without external intervention, although the severity of disease and time course for healing may vary enormously depending on the species of infecting parasite as well as the nature of the immune response by an infected individual. Current estimates suggest a global annual incidence of 1–1.5 cases of CL, most of which occur in Afghanistan, Iran, Saudi Arabia, and Syria in the Old World and Brazil and Peru in the New World (1). Since most species of *Leishmania* causing human disease also infect mice, murine cutaneous leishmaniasis has been extensively studied as a model for understanding the immune regulation of this intracellular infection. This chapter will summarize how knowledge of experimental murine infections has impacted our understanding of human disease.

EXPERIMENTAL CUTANEOUS LEISHMANIASIS
Leishmania major in mice

Although several species of *Leishmania* infect mice and produce disease similar to that seen in humans, *L. major* infection is undoubtedly the most the intensively studied. It was noted in the early 1980s that *L. major* infection in most strains of inbred mice resulted in the development of small, self-healing cutaneous lesions while infection in BALB/c mice led to the development of large, non-healing cutaneous ulcers (2,3). Classical studies by James Howard and F.Y. Liew, also in the early 1980s, showed that the exaggerated susceptibility of BALB/c mice to *L. major* could be reversed if the mice were sublethally irradiated prior to infection (4). In addition, these investigators showed that the cells associated with resistance and those associated with progressive disease were both CD4+ T cells (5). At the time, the CD4+ cells associated with susceptibility were defined as "suppressor" T cells, although it is now clear that these studies actually identified a population of Th2 effector cells. Shortly after the initial description of CD4+

Th1 and Th2 cells came evidence that healing and non-healing infections with *L. major* correlated with the development of dominant Th1 and Th2 type responses, respectively (6-8). These early studies formed the basis for future research using *L. major* infected mice as a model for understanding the in vivo regulation of CD4+ subsets.

It is now clear that the association between Th1 cells and resistance resides in their ability to produce the macrophage-activating cytokine, IFNγ. Treatment of resistant mice with neutralizing antibody to IFNγ (9) exacerbates infection and mice with disruptions of either the IFNγ gene (10) or the *IFN*γ receptor gene (11) are unable to control infection. The dominant role of IFNγ during infection is to activate macrophages to kill intracellular parasites. Among the products produced by activated macrophages, nitric oxide (NO) has emerged as the dominant oxidant mediating killing of leishmanial parasites in mice. NO is produced via the oxidation of L-arginine by the enzyme, iNOS or inducible NO synthase (12). IFNγ, alone, is capable of enhancing the production of iNOS by macrophages but maximum NO production occurs following stimulation of macrophages with IFNγ and a second cytokine, TNFα (13). That NO is directly involved in parasite killing is shown by observations that specific inhibitors of NO production block killing of *Leishmania* in vitro by activated macrophages while in vivo treatment with NO inhibitors reverses the ability of resistant mice to control infection (12,14). In addition, administration of NO inhibitors to C57BL/6 mice that have healed an infection with *L. major* can lead to reactivation of disease (15).

Development and maintenance of polarized Th1/Th2 responses

Numerous studies have examined the early event during infection that contribute to the activation of dominant Th1 or Th2 responses. In susceptible BALB/c mice, treatment with recombinant IL-12 or neutralization of IL-4 can block the development of a Th2 response and result in the activation of a protective Th1 response following *L. major* infection (16-18). Conversely, treatment of resistant strains of mice with antibodies to IL-12 or IFNγ can exacerbate disease and promote the development of a Th2 response (19,20). Such treatments must be initiated either prior to or near the time of parasite inoculation in order to be effective suggesting that they influence the early events in the differentiation of naïve T cells into Th1 or Th2 effector cells. It is now clear that T cells expressing identical TCR can develop into both Th1 and Th2 cells, suggesting that the nature of a specific antigen may not be the critical factor in the induction of specific Th subsets. Instead, most evidence suggests that the cytokine milieu present during T cell activation is extremely important in determining the pathway of Th subset development (21,22). These results would predict that events that regulate the development of

Th1/Th2 responses occur rapidly following infection. Analysis of early cytokine responses in non-healing BALB/c mice shows an increase in IL-4 production within draining lymph nodes during the first 24 hours of infection (23). The IL-4 appears to be produced by a population of CD4+ T cells that express Vβ4-Vα8 T cell receptors that recognize the *Leishmania* homolog of mammalian RACK1 termed LACK or *Leishmania*-activated C kinase (24,25). In contrast to the response seen in BALB/c mice, draining LN cells from resistant C3H mice produce high levels of IFNγ but little IL-4, following infection (26). NK cells seem to be a major source of IFNγ production on day 2-3 of infection in these mice (26). Although these and other studies have revealed clear differences in early cytokine production, they have also exposed similarities in the responses in resistant and susceptible mice. Careful comparisons of cytokine production within lymph nodes draining local sites of infection have revealed that both IL-4 and IFNγ are produced in both resistant C57BL/6 and susceptible BALB/c mice during the first few days after parasite inoculation, although IL-4 production was shown to be higher in BALB/c mice (27). Following an initial peak in IL-4 gene transcription, IL-4 mRNA levels are down regulated in the healer strain whereas IL-4 transcription increases in non-healing BALB/c mice (27).

As previously noted, BALB/c mice can be induced to heal an infection by manipulating the immune response prior to parasite inoculation. In addition to treatment of mice with sublethal irradiation (4) or antibodies to IL-4(28),treatment with antibodies to CD4, IL-2, or TGFß will induce healing in infected mice (18, 29-31). In addition, treatment of mice with CTLA4-Ig (32) that blocks CD28 costimulation will promote healing in BALB/c mice. Given the known importance of IL-4 in promoting Th2 development, it is likely that some of these treatments suppress early IL-4 production, either specifically, as is the case with anti-IL-4 treatment, or by non-specifically reducing the magnitude of the initial immune response as would occur following in vivo treatment with antibodies to CD4 or IL-2 (28,29). IL-12 treatment, in contrast, may function to directly stimulate IFNγ production and Th1 development (reviewed in 33). In some resistant strains of mice such as the C3H, IL-12 may stimulate NK cells to produce an early burst of IFNγ production which, in turn, plays a role in promoting a Th1-type response (26). IL-12 has also been shown to suppress IL-4 production and may do so via an IFNγ-independent mechanism (34). The production of IL-12 in vivo has been suggested to correlate with infection of macrophages by the amastigote stage of *L. major* (27). Supporting the role of IL-12 in promoting Th1 cell development, in vivo treatment with an antibody to IL-12 has been shown to reverse the healing in resistant C57BL/6 mice (16,35). The importance of IL-12 production has recently been emphasized in studies showing that in the absence of continued IL-12 production, Th1 type responses may wane and Th2 type responses emerge in BALB/c mice induced to heal by IL-12 therapy administered early during infection (36,37). Whether IL-12 acts directly to maintain an established Th1 cell population or functions to prevent the development of a Th2 type response is yet to be clarified.

Additional insight into the factors which regulate healing and non-healing in infected mice has come from the studies by Peter Bretscher, which has shown that the inoculation of low numbers of L. major promastigotes (1-3 x 10^2) into susceptible BALB/c mice leads to a stable Th1 response whereas inoculation of higher numbers of promastigotes (> 10^3) results in the activation of a Th2-type response and progressive disease (38). It has been suggested that the threshold for Th2 activation may be lower in BALB/c mice than in healing strains and that a stable CMI response is generated in this strain only by inoculation of very low numbers of parasites (38). This hypothesis is consistent with earlier studies demonstrating "immune deviation" in which immunization with low amounts of antigen elicited stable DTH responses and low antibody production, whereas higher antigen doses induced antibody production, but no DTH reactivity (39,40). Once DTH or antibody production had been elicited in immunized mice, it was nearly impossible to induce the reciprocal response in individual animals, suggesting that under appropriate conditions, immunization can "lock in" a CMI versus humoral response (40). The initial cytokine response to low dose Leishmania infections is difficult to analyze, however, it might be predicted that low parasite inocula may not elicit the early burst in IL-4 production required for Th2 development. However, low parasite inocula do not always induce a healing response since recent studies have shown that BALB/c mice inoculated with very low numbers of metacyclic promastigotes may also develop dominant Th2 type responses and fail to control infection (41,42), suggesting that strains of L. major may differ in virulence.

During the past few years, our knowledge of how costimulatory molecules function to regulate T cell activation has increased considerably. As note above, treatment of BALB/c mice with CTLA4Ig, which blocks interactions between B7.1/B7.2 on APCs and both CD28 and CTLA-4 expressed on T cells, will promote healing if given prior to infection, but block healing if administered continually throughout infection. Curiously, given these results, CD28-/- mice on either the BALB/c or C57BL/6 background do not exhibit altered patterns of infection or cytokine production suggesting that CD28-B7 interactions are not absolutely required for the development of Th2 versus Th1 type responses (43). However, these studies used a high parasite inoculum and different results were obtained when CD28-/- mice were challenged with only a few hundred parasites. In these more recent studies, BALB/c mice lacking the gene for CD28 developed a

Th1 type response and controlled infection suggesting that, in certain circumstances, cositmulation, may influence not only the strength of an immune responses but also the phenotype of that response (42).

In contrast, studies of another costimulatory interaction, namely CD40-CD40L, have revealed an extremely important role for these molecules in the generation of a Th1 type response. Mice genetically deficient in either CD40 or CD40L, or mice treated with neutralizing antibodies to these molecules, are highly susceptible to infection with both *L. major* and *L. amazonensis* (44-46). Analysis of the immune response in these mice reveals reduced levels of IFNγ production that could be reversed by treatment of mice with recombinant IL-12. Since stimulation through CD40 on macrophages and dendritic cells by CD40L on T cells promotes IL-12 production, it seems logical that a major pathway for IL-12 production during leishmanial infections involves CD40-CD40L costimulation. However, it is still unclear whether CD40-CD40L costimulation is necessary for the production of an early burst of IL-12 production required for the induction of a Th1 type response or whether this costimulatory pathway plays a more important role in sustaining IL-12 production required for maintenance of a Th1 response.

Although disruption of the CD40-CD40L costimulatory pathway reduces IL-12 production and reverses the normal healing pattern of an *L. major* infection in a resistant strain of mouse, stimulation through CD40 may not be the only mechanism responsible for IL-12 production. Toll-like receptors (TLR) on antigen presenting cells can recognize pathogen-associated pattern recognition molecules such as bacterial LPS, peptoglycans and lipoproteins as well as glycosylphosphadylinositol anchors of *Trypanosoma cruzi* (reviewed in 47). Although direct infection of macrophages by *Leishmania* does not appear to elicit IL-12 production, recent studies have shown that dendritic cells infected in vitro with *L. major* or *L. donovani* do produce IL-12 (48,49). In addition, it has been suggested that a protein from *L. major*, termed Leif, can directly stimulate IL-12 production by human monocytes (50). As noted above, it has also been suggested that infection of macrophages by amastigotes, rather than by promastigotes, may induce IL-12 production. Whether any of these parasite-host cell interactions that induce IL-12 production involve TLRs is currently unclear. However, we have recently shown that mice lacking the genes for both CD40L and CD28 produce detectible levels of IL-12 early during *L. major* infection and can resolve their cutaneous lesions (51). In addition, following inoculation with low numbers of promastigotes, CD40L deficient mice can also control infection (51). Thus, it is likely that interactions other that CD40-CD40L can provide a stimulus for IL-12 production.

Although CD4+ Th1 cells play a dominant role in resistance to *L. major*, CD8+ cells are also activated during infection may be an important component in the response to reinfection. Depletion of CD8+ cells in genetically resistant CBA mice following healing can exacerbate lesion development during a challenge infection with *L. major* (52). Treatment with anti-CD8 antibody also reverses resistance to challenge in susceptible BALB/c mice induced to heal their primary infection following administration of anti-IL-4 antibody (52). CD8+ T cells contribute significantly to the production of IFNγ in these healed and rechallenged mice. In addition, depletion of CD8+ cells in mice immunized with irradiated promastigotes or vaccinated with DNA encoding the leishmanial LACK antigen, blocks resistance to reinfection (53,54). In contrast, in vivo depletion of CD8+ cells has only a marginal effect of the primary course of infection in CBA mice and mice genetically deficient in β2-microglobulin or the CD8α gene, both of which lack functional CD8+ T cells, can control a primary infection with *L. major*, and in the case of CD8α-deficient mice, exhibit resistance to reinfection (55,56). It is probable that CD8+ T cells are a normal component of the immune response both vaccinated and infected mice, but that CD4+ cells can fully compensate for their absence during a primary infection.

Even though much is known about the events governing differentiation of naïve CD4+ T cells into Th1 or Th2 cells, relatively few studies have addressed how established Th2 type responses may be switched to protective Th1 responses. As noted above, most in vivo treatments with cytokines or anti-cytokine antibodies that promote healing in *L. major*-infected BALB/c mice must be initiated at of near the time of parasite inoculation. However, we have shown that BALB/c mice treated with IL-12, IFNγ or anti-IL-4 antibody after several weeks of infection will resolve their lesions and develop Th1 type responses if concurrently treated with an anti-leishmanial drug such as sodium stibogluconate (57-59). In addition, prolonged treatment of BALB/c mice with sodium stibogluconate for a period of 3 months starting at week 3 of infection promoted cure and a Th1 type response in a majority of the animals (60). Since drug treatment reduces parasite numbers as well as limits the accumulation of inflammatory macrophages at the site of infection, it is possible that drug treatment may exert an effect on the production of immunosuppressive cytokines by inflammatory macrophages. For example, transcript levels for TGFβ in parasitized lesions are reduced following drug therapy of BALB/c mice (59). In addition, treatment with anti-TGFβ antibody promotes healing in chronically infected CB6F1 mice (61). More recently, it was shown that immunotherapy with DNA encoding a leishmanial surface antigen (PSA-2) can modulate lesion expansion and promote a shift toward a Th1 type response in infected BALB/c mice (62). Whether these treatments promote the activation of naïve T cells into Th1 cells or promote the expansion of preexisting populations of Th1 in chronically infected mice is currently unclear. Another regulatory cytokine produced by macrophages and T cells is IL-10. Recent studies show that IL-10 deficient BALB/c can control infection with *L. major* (63). In addition, sterile cure can be achieved in IL-10 deficient C57BL/10 mice suggesting that IL-10 plays a key role in susceptibility to infection as well as maintenance of parasite persistence following lesion resolution (64).

Experimental studies with other *Leishmania* species.

Although *L. major* has been the most intensively studied *Leishmania* species in mice, the immune response to other species has not been neglected and studies of New World species, in particular, offer interesting contrast to those with *L. major*. For example, *Leishmania braziliensis* infection in BALB/c mice results in the development of small, self-healing lesions that contrast to those produced by *L. major* (65). Although there is limited data on cytokine production during *L. braziliensis* infection, treatment of mice with antibody to IFNγ markedly enhances lesion size suggesting the BALB/c mice may develop a protective Th1 type response (65). In contrast to the apparent low virulence of *L.* braziliensis parasites of the *L. mexicana* complex produce disease that is at least as severe as that caused by *L. major*. For example, both *L. major* and *L. amazonensis* produce progressive, non-healing infections in BALB/c mice (66,67). However, *L. major* infections heal in C57BL/10 mice while *L. amazonensis* infection results in the development of small bur persistent lesions. (68). C57BL/6 mice also fail to resolve infection with *L. amazonensis* and exhibit a defect in the production of IFNγ rather than the production of high levels of IL-4 (69). C3H mice that are resistant to *L. major* develop a similar pattern of chronic infection with *L. amazonensis*. Interestingly, cells from *L. amazonensis*-infected mice produced low levels of IL-12, but were not induced to heal following treatment with exogenous IL-12 (69). In addition, IL-4 deficient mice were no more resistant to infection that wild type mice suggesting that the failure to resolve infection was independent of the development of a dominant Th2 type response (69). The underlying mechanism responsible for the failure of resistant strains of mice to develop a vigorous Th1 type response following infection with *L. amazonensis* is unclear, but could involve the production of anti-inflammatory factors such as TGFβ which has been shown to play a role in the susceptibility of BALB/c mice to this parasite (30).

The role of T cells in the pathogenesis of disease seems to differ in mice infected with *L. major* and *L. amazonensis*. While MHC II-deficient mice which lack functional CD4+ T cells exhibit a normal course of lesion development during the first 5 weeks of infection with *L. major*, similar mice fail to develop detectable lesions over 12 weeks following infection with *L.*

amazonensis (70). However, reconstitution of T cell deficient mice with naïve CD4+ cells restores their capacity to develop lesions suggesting that CD+ T cells are critical for recruitment of inflammatory macrophages into lesions that, in turn, are required to support parasite replication. Interestingly, the disease-promoting CD4+ T cells in *L. amazonensis*-infected mice have the characteristics of Th1 cells. (70).

Another interesting contrast betweens infections with *L. major* and *L. mexicana* involves the role of antibody during infection. Neither BALB/c nor C57BL/6 mice in which the IgM locus (μMT mice) has been deleted exhibit altered patterns of cutaneous disease following infection with *L. major*. However, lesion development is significantly delayed in mMT mice infected with either *L. mexicana* or *L. pifanoi* (71). A similar defect in lesion development is observed following infection of Fc receptor deficient mice with these New World parasites suggesting that immunoglobulin-mediated uptake of parasites by macrophages is critical to sustaining infection with *L. mexicana* but *L. major* (71) Whether antibody-mediated opsonization of these Old and New World parasites leads to different patterns of cytokine production by infected macrophages is currently unclear. However, it does seem probable from these studies and those mentioned above that the Th1/Th2 paradigm that describes susceptibility and resistance to *L. major* in mice does not necessarily apply to infection with parasites of the *L. mexicana* complex. Rather, susceptibility to *L. mexicana* and *L. amazonensis* correlates with a reduced Th1 type response rather that an exaggerated Th2 response, suggesting that parasite factors, as well as the host response, are important components in the development of resistance.

HUMAN CUTANEOUS LEISHMANIASIS

In general, cutaneous infection with most dermotropic species of *Leishmania* results in the development of mild to severe ulcerating lesions that spontaneously resolve over a variable period of time. Occasionally, infected individuals develop an abnormal form of disease termed diffuse cutaneous leishmaniasis (DCL) in which the architecture of non-healing dermal lesions is dominated by the presence of parasitized macrophages and a paucity of lymphocytes. These polar extremes in clinical presentation have contributed to the concept that leishmaniasis is a "spectral" disease determined by the type of immune response which develops during infection. Still, abnormal clinical presentations are associated with specific *Leishmania* species. For example, *L amazonensis* and *L. aethiopica*, both of which cause simple cutaneous infection, are responsible for most cases of DCL while lupoid leishmaniasis, a chronic form of disease associated with the development of papules or nodules around a healing primary lesion, is primarily seen in patients infected with *L. tropica*. Mucosal leishmaniasis is generally associated with *L. braziliensis*, but occasionally is seen in patients with *L. panamensis* or more rarely, *L. guyanensis* (72).

As might be predicted from murine studies, there is considerable evidence that self-healing human leishmaniasis is associated with the development of a Th1 type response. For example, patients from an area endemic for *L. braziliensis* who experienced rapidly healing infections exhibited low levels of anti-*Leishmania* antibody, but strong delayed type skin tests following inoculation of leishmanial antigen (73). In addition, stimulation of peripheral blood cells from these individuals elicited the production of IFNγ (73). Interesting, the parameters associated with cell-mediated immunity (DTH, IFNγ production) were higher in individuals who had resolved infection compared to those with active disease (73). Some additional insight into how responses are regulated during infection comes from a study of individuals infected with *L. braziliensis*. During early stages of infection 9< 2 months), both proliferative responses and IFNγ production were low compared to responses in individuals at later stages (> 2 months) of infection (74). In contrast, IL-10 production was high in early stages of infection but decreased as lesions resolved following chemotherapy. IFNγ production could be restored in cultures of cells from non-responding patients by neutralization of IL-10 or by addition of IL-12 (74). These results suggest that IL-10 may function during early stages of infection to down-regulate IL-12 and IFNγ production, thereby allowing parasites to dramatically increase in number. IL-13 has also been implicated as a cytokine capable of modulating the development of resistance. IL-13 mRNA levels were shown to be elevated in active lesions due to infection with L. guyanensis, but reduced following healing. The presence of elevated levels of IL-13 in lesions correlated with reduced expression of the β2 subunit of the IL-12 receptor and neutralization of IL-13 in cultures of patient cells increased expression of the IL-12β2R as well as IFNγ production (75). Slightly different results were observed in studies of individuals infected with *L. major*. While IFNγ production correlated with resistance to infection, no differences in IFNγ production by PBLs was noted in patients with active, healing infections and those who had resolved infection (76). However, cells from individuals with chronic, non-healing infections exhibited a distinct response to stimulation with parasite antigen with decreased IFNγ production and increased production of IL-4 (76). CD4+ T cells were the major source of both IFNγ and IL-4 production suggesting that dichotomous Th1 and Th2 type responses are associated with the ability or inability to control human *L. major* infection (76).

Although IFNγ-producing CD4+ T cells are clearly associated with resistance to cutaneous infection, this is also considerable evidence that CD8+ T cells play a role in healing. Analysis of responses by PBLs from *L.*

braziliensis patients before and after antimony chemotherapy showed an increase in the percentage of CD8+ and decrease in CD4+ cells responding to leishmanial antigen (77). A more recent study confirmed the increase in CD8+ responsive cells following therapy and showed that both CD4 and CD8+ cells were a source of IFNγ production (77). An additional study of patients with lesions of less that two months duration living in an area endemic for *L. braziliensis* confirmed that both CD4+ and CD8+ cells contribute significantly to IFNγ production and demonstrated that neither of these cell populations produced IL-4, IL-5, or IL-10 (79).

Although it is well established that activated murine macrophages can kill *Leishmania* via production of the reactive nitrogen intermediate, nitric oxide (NO), a similar pathway in human cells has been more difficult to demonstrate. However, several studies now implicate NO as the mediator for intracellular killing in human cells. Stimulation of human peripheral mononuclear cells with IFNγ or by cross linking surface CD23/ FcεRII can stimulate the production of NO as assessed by the generation of nitrate, the stable oxygenation product of NO. Such activated cells could kill *L. major* in vitro (80). More recently, the capacity of human monocytes to kill *Leishmania* via an NO dependent mechanism was confirmed. Increased levels of iNOS mRNA and protein could be detected following infection of IFNγ stimulated cells. Although actual NO production by IFNγ-stimulated cells could not be observed, the ability of activated cells to kill intracellular amastigotes was reversed in cultures containing the iNOS inhibitor, N^G-monomethyl-L-arginine (reviewed in 81).

If activation of IFNγ-producing CD4+ and CD8+ T cells correlates with healing and the development of resistance to infection, what accounts for the failure of some cutaneous lesions to heal in a timely manner, if at all? The answer may come from studies of immunological responses in patients with DCL. As opposed to healing lesions that show the presence of infiltration by CD4+ and CD8+ T cells, lesions from patients with DCL generally show a predominance of infected macrophages with few T cells. A DTH response to leishmanial antigens is absent, but parasite-specific antibody levels are high (82,83). Although diffuse cases are rare, analysis of immune responses in patients with DCL are consistent with a Th2 type profile of cytokine production. Direct measurements of cytokine mRNA levels in cutaneous lesions of Brazilian patients consistently show that transcripts for IFNγ are low while those for Th2 type cytokines such as IL-4, IL-5, and IL-10 are elevated (84,85). Elevated levels of IL-10 are also seen in patients with diffuse leishmaniasis due to infection with *L. aethiopica* 86). Levels of IgG4, an antibody isotype associated with Th2 type responses, are also significantly elevated in individuals with DCL (87). Cytokines produced by PBLs from individuals with DCL also exhibit a general Th2 type bias (88).

Interestingly, cells from DCL patients undergoing chemotherapy often produce increased amounts of IFNγ and decreased amounts of Th2 type cytokines such as IL-10, although this shift toward Th1 dominance is generally insufficient to prevent relapse following the termination of therapy (88). Whether this means that other factors in addition to the balance of Th1 to Th2 type cytokines play a role in the pathogenesis of DCL is currently unclear. For example, TGFβ has been detected in lesions during early stages of cutaneous infection as well as in mucosal lesions (89). In addition, high levels of mRNA for TGFβ were detected in lesions of greater that 4 months duration (90), although TGFβ is post-transcriptionally regulated and high message levels may not necessarily translate to high levels of bioactive TGFβ protein.

Although localized cutaneous leishmaniasis and DCL are polar extremes, marked variations may occur in the severity and time to healing of cutaneous infections with most species of *Leishmania*. It is logical to think that variations in the immune response among individuals will affect the clinical outcome. Indeed, as mentioned above, individuals with either acute *L. major* infections versus chronic, non-healing *L. major* infections showed a dichotomy in IFNγ versus IL-4 production by peripheral CD4+ T cells (76). Thus, the balance of pro-inflammatory cytokines associated with macrophage activation (IFNγ and TNFα) versus cytokines that can suppress macrophage activation or moderate the development of a Th1 type response (IL-4, IL-10, or TGFβ) seems to predict the outcome of most cutaneous infections. However, the immune response to infection does not adequately explain the pathogenesis of mucocutaneous leishmaniasis (MCL). MCL involving the mucosa and cartilage of the upper respiratory tract often occurs years after healing of a primary dermal lesion. Typically, MCL patients exhibit all of the parameters of a positive cell-mediated response. Their cells proliferate better and produce more IFNγ than cells from patients with simple cutaneous lesions in response to leishmanial antigen, although this may have to do with the duration of infection rather than the pathogenesis of mucosal disease (82,83,91). Analysis of cytokine mRNA levels in lesions of MCL patients suggests a mixed Th1/Th2 response (84). Message for IFNγ , IL-4 and TNFα was present in lesions of most MCL patients as well as patients with localized lesions , but mRNA for IL-10 was more abundant in MCL lesions (84). An examination of the numbers of CD4+ T cells expressing a "memory" phenotype or producing IFNγ appears to be similar in individuals with localized versus mucosal lesions (85). Levels of IgG isotypes in the sera of MCL and LCL patients are also consistent with the presence of a dominant Th1 type response during both forms of disease (87). The failure of define distinct features of the immune response to explain the chronic, destructive nature of MCL suggests that unique parasite determinants play an important role in the development of mucosal disease. Although mucosal lesions can be caused by multiple species of *Leishmania*, they are most common following infection with *L. braziliensis*, a parasite known to produce chronic, progressive lesions. Since parasites are generally sparse in mucosal lesions, it is likely that the same immune response that controls infection at dermal sites is responsible for tissue destruction at mucosal sites. Interesting insight into the pathogenesis of DCL comes from studies linking specific alleles of the TNFα and TNFβ genes with an increases frequency of mucosal disease. A 7.5-fold higher risk of MCL occurs in individuals homozygous for an allelic polymorphism in the gene for TNFβ and a 3.5-fold increased risk for MCL associated with a single nucleotide polymorphism in the gene for TNFα (92). These observations suggest that genetic polymorphisms associated with TNF regulatory elements may contribute to the high levels of circulating TNFα observed in DCL patients (reviewed in 93). Although TNFα is critical to parasite killing, excessive, or unregulated production of this cytokine could clearly contribute to the immunopathology seen in DCL lesions.

CONCLUSIONS

If an understanding of the immune response to any one disease has benefited from experimental studies in mice, that disease is cutaneous leishmaniasis. The popularity of rodent infection models, combined with the fact that many *Leishmania* species produce similar infections in mice and humans, has paved the way for studies which are now uncovering the mechanisms by which altered immunological responses influence the clinical picture of human infection. We now know that Th1 type responses are required for control of infection in both mice and humans. There is also significant evidence that cytokines such as IL-10 and TGFβ that negatively influence resistance in mice may also play a role in the development of persistent infections in humans. What remains unclear is why individual responses to infection are so varied. Still, even without a clear understanding of why the same parasite can produce clinically disparate symptoms in certain patients, we now have sufficient knowledge to proceed with the development of effective vaccines and immunotheraputic treatments for human use.

REFERENCES

1. Desjeux P. 2001. The increase in risk factors for leishmaniasis worldwide. Trans R Soc Trop Med Hyg. 95: 239-243.
2. Howard JG, Hale C, and Chan-Liew WL. 1980. Immunological regulation of experimental cutaneous leishmaniasis. I. Immunogenetic aspects of susceptibility to *Leishmania tropica* in mice. Parasite Immunol 2:303-314.
3. Detolla LJ,Jr., Scott PA, and Farrell JP. 1981. Single gene control of resistance to cutaneous leishmaniasis in mice. Immunogenetics 14:29-39.

4. Howard JG, Hale C, and Liew FY. 1981 Immunological regulation of experimental cutaneous leishmaniasis. IV. Prophylactic effect of sublethal irradiation as a result of abrogation of suppressor T cell generation in mice genetically susceptible to Leishmania tropica. J Exp Med. 153:557-68.

5. Liew FY, Hale C, and Howard JG. 1982 Immunologic regulation of experimental cutaneous leishmaniasis. V. Characterization of effector and specific suppressor T cells. J Immunol. 128:1917-22.

6. Locksley RM, Heinzel FP, Sadick MD, Holaday BJ, and Gardner KD Jr. 1987 Murine cutaneous leishmaniasis: susceptibility correlates with differential expansion of helper T-cell subsets. Ann Inst Pasteur Immunol 138:744-749.

7. Scott P, Natovitz P, Coffman RL, Pearce E, and Sher A. 1988. Immunoregulation of cutaneous leishmaniasis: T cell lines that transfer protective immunity or exacerbation belong to different T helper subsets and respond to distinct parasite antigens. J Exp Med 168:1675-1684.

8. Heinzel FP, Sadick MD, Holaday BJ, Coffman RL, and Locksley RM. 1989.Reciprocal expression of interferon γ or interleukin 4 during the resolution or progression of murine leishmaniasis. J Exp Med 169:59-72.

9. Belosevic M, Finbloom DS, Van Der Meide PH, Slayter MV, and Nacy CA. 1989. Administration of monoclonal anti-IFN-γ antibodies in vivo abrogates natural resistance of C3H/HeN mice to infection with *Leishmania major*. J Immunol 143:266-274.

10. Wang Z-E, Reiner SL, Zheng S, Dalton DK, and Locksley RM. 1994. CD4+ effector cells default to the Th2 pathway in interferon-g -deficient mice infected with *Leishmania major*. J Exp Med 179:1367-1371.

11. Swihart K, Fruth U, Messmer N, Hug K, Behin R, Huang S, Del Giudice G, Aguet M, and Louis JA. 1995. Mice from a genetically resistant background lacking the interferon gamma receptor are susceptible to infection with *Leishmania major* but mount a polarized T helper cell 1-type CD4+ T cell response. J Exp Med. 181:961-971.

12. Liew FY, Millott S, Parkinson C, Palmer RM, and Moncada S. 1990 Macrophage killing of Leishmania parasite in vivo is mediated by nitric oxide from L-arginine. J Immunol; 144:4794-4797.

13. Liew FY, Li Y, and Millott S. 1990 Tumor necrosis factor-alpha synergizes with IFN-gamma in mediating killing of Leishmania major through the induction of nitric oxide. J Immunol. 145:4306-4310.

14. Evans TG, Thai L, Granger DL, and Hibbs JB Jr. 1993. Effect of in vivo inhibition of nitric oxide production in murine leishmaniasis. J Immunol. 151:907-915.

15. Stenger S, Donhauser N, Thuring H, Rollinghoff M, and Bogdan C. 1996. Reactivation of latent leishmaniasis by inhibition of inducible nitric oxide synthase. J Exp Med. 183:1501-1514.

16. Sypek JP, Chung CL, Mayor SEH, Subramanyam JM, Goldman SJ, Sieburth DS, Wolf SF, and Schaub RG. 1993. Resolution of cutaneous leishmaniasis: interleukin 12 initiates a protective T helper type 1 immune response. J Exp Med 177:1797-1802.

17. Heinzel FP, Schoenhaut DS, Rerko RM, Rosser LE, and Gately MK. 1993. Recombinant interleukin 12 cures mice infected with *Leishmania major*. J Exp Med 177:1505-1509.

18. Sadick MD, Heinzel FP, Holaday BJ, Pu RT, Dawkins RS, and Locksley RM. 1990. Cure of murine leishmaniasis with anti-interleukin 4 monoclonal antibody. Evidence for a T cell-dependent, interferon gamma-independent mechanism. J Exp Med 171:115-125.

19. Scharton-Kersten T, Afonso LCC, Wysocka M, Trinchieri G, and Scott P. 1995. IL-12 is required for natural killer cell activation and subsequent T helper 1 cell development in experimental leishmaniasis. J Immunol 154:5320-5330.

20. Scott P. 1991. IFN-gamma modulates the early development of Th1 and Th2 responses in a murine model of cutaneous leishmaniasis. J Immunol. 147:3149-3155.

21. Hsieh CS, Heimberger AB, Gold JS, O'Garra A, and Murphy KM. 1992. Differential

regulation of T helper phenotype development by interleukins 4 and 10 in an alpha beta T cell receptor transgenic system. Proc Nat Acad Sci U S A 89:6065-6069.

22. Hsieh CS, Macatonia SE, Tripp CS, Wolf SF, O'Garra A, and Murphy KM. 1993. Development of Th1 CD4+ T cells through IL-12 produced by *Listeria*-induced macrophages. Science 260:547-549.

23. Himmelrich H, Parra-Lopez C, Tacchini-Cottier F, Louis JA, and Launois P. 1998. The IL-4 rapidly produced in BALB/c mice after infection with Leishmania major down-regulates IL-12 receptor beta 2-chain expression on CD4+ T cells resulting in a state of unresponsiveness to IL-12. J.Immunol.161:6156-6163.

24. Reiner SL, Wang ZE, Hatam F, Scott P, and Locksley RM. 1993. TH1 and TH2 cell antigen receptors in experimental leishmaniasis. Science. 259:1457-1460.

25. Himmelrich H, Launois P, Maillard I, Biedermann T, Tacchini-Cottier F, Locksley RM, Rocken M, and Louis JA. 2000. In BALB/c mice, IL-4 production during the initial phase of infection with Leishmania major is necessary and sufficient to instruct Th2 cell development resulting in progressive disease. J Immunol. 164:4819-4825.

26. Scharton TM and Scott P. 1993. Natural killer cells are a source of interferon gamma that drives differentiation of CD4+ T cell subsets and induces early resistance to *Leishmania major* in mice. J Exp Med 178:567-577.

27. Reiner SL, Zheng S, Wang Z, Stowring L, and Locksley RM. 1994. *Leishmania* promastigotes evade interleukin 12 (IL-12) induction by macrophages and stimulate a broad range of cytokines from CD4+ T cells during initiation of infection. J Exp Med 179:447-456.

28. Titus RG, Ceredig R, Cerrottini JC, and Louis J. 1985. Therapeutic effect of anti-L3T4 monoclonal GK1.5 on cutaneous leishmaniasis in genetically susceptible BALB/c mice. J Immunol 135:2108-2114.

29. Heinzel FP, Rerko RM, Hatam F, and Locksley RM. 1993. Interleukin 2 is necessary for progression of leishmaniasis in susceptible murine hosts. J Immunol 150:3924-3931.

30. Barral-Netto M, Barral A, Brownell CE, Skeiky YA, Ellingsworth LR, Twardzik DR, and Reed SG. 1992. Transforming growth factor-beta in leishmanial infection: a parasite escape mechanism. Science 257:545-548.

31. Corry DB,. Reiner, SL, Linsley PS and Locksley. RM. 1994. Differential effects of blockade of CD28-B7on the development of Th1 or Th2 effector cells in experimental leishmaniasis. J Immunol 153:41424148.

32. Reiner SLand Locksley RM. 1995. The regulation of immunity to *Leishmania major*. Ann Rev Immunol 13:151-177.

33. Trinchieri G. 1995. Interleukin 12: a proinflammatory cytokine with immunoregulatory functions that bridge innate resistance and antigen-specific adaptive immunity. Ann Rev Immunol 13:251-276.

34. Wang Z-E, Zheng S, Corry DB, Dalton DK, Seder RA, Reiner SL, and Locksley RM. 1994. Interferon-γ -independent effects of interleukin 12 administered during acute or established infection due to *Leishmania major*. Proc Nat Acad Sci U S A 91: 12932-12936.

35. Constantinescu CS, Hondowicz BD, Elloso MM, Wysocka M, Trinchieri G, and Scott P. 1998. The role of IL-12 in the maintenance of an established Th1 immune response in experimental leishmaniasis. Eur J Immunol. 28:2227-2233.

36. Park AY, Hondowicz BD, and Scott P. 2000. IL-12 is required to maintain a Th1 response during Leishmania major infection. J Immunol. 165:896-902.

37. Stobie L, Gurunathan S, Prussin C, Sacks DL, Glaichenhaus N, Wu CY, and Seder RA. 2000. The role of antigen and IL-12 in sustaining Th1 memory cells in vivo: IL-12 is required to maintain memory/effector Th1 cells sufficient to mediate protection to an infectious parasite challenge. Proc Natl Acad Sci U S A. 2000 978427-8432.

38. Bretscher PA, Wei G, Menon JN, and Bielefeldt-Ohmann H. 1992. Establishment of stable, cell-mediated immunity that makes "susceptible" mice resistant to *Leishmania major*. Science 257:539-542.

39. Lagrange PH, MacKaness GB, and Miller TE. 1974. Influence of dose and route of antigen injection on the immunological induction of T cells. J Exp Med 139:528-542.

40. Parish CR. 1972. The relationship between humoral and cell-mediated immunity. Transplant Rev 13:35-66.

41. Belkaid Y, Kamhawi S, Modi G, Valenzuela J, Noben-Trauth N, Rowton E, Ribeiro J, and Sacks DL. 1998. Development of a natural model of cutaneous leishmaniasis: powerful effects of vector saliva and saliva preexposure on the long-term outcome of *Leishmania major* infection in the mouse ear dermis. J Exp Med 188:1941-1953.

42. Compton HL and Farrell, JP 2002. CD28 costimulation and parasite dose combine to influence the susceptibility of BALB/c mice to infection with *Leishmania major*. J. Immunol.168:1302-1308.

43. Brown DR, Green JM, Moskowitz, NHM, Thompson CB and. Reiner SL 1996. Limited role of CD28-mediated signals in T helper subset differentiation. J Exp Med 184:803-810.

44.Ferlin WG., von der Weid T, Cottrez F, Ferrick DA, CoffmanRL and Howard MC. 1998. The induction of a protective response in *Leishmania major*- infected BALB/C mice with anti-CD40 mAb. Eur J Immunol.2:525-531.

45.Cambell KA, Ovendale PJ, Kennedy MK, Fanslow WC, Reed SG and Maliszewski. CR. 1996. CD40 ligand is required for protective cell-mediated immunity to *Leishmania major*. Immunity 4:283-289.

46. Soong, L, Xu JC, Grewal IS, Kima P. Sun J, Longley BJ.Jr, Ruddle NH, McMahon-Pratt D and Flavell RA. 1996. Disruption of CD40-CD40 ligand interactions results in an enhanced susceptibility to *Leishmania amazonensis* infection. Immunity 4:263-273.

47. Akira S, Takeda K, Kaisho T. 2001 Toll-like receptors: critical proteins linking innate and acquired immunity. Nat Immunol. 2:675-680.

48. Konecny P, Stagg AJ, Jebbari H, English N, Davidson RN and Knight SC. 1999. Murine dendritic cells internalize *Leishmania major* promastigotes, produce IL-12 p40 and stimulate primary T cell proliferation in vitro. Eur. J. Immunol. 29:1803-1811.

49. Flohe S.B, Bauer C, Flohe S, and Moll H. 1998. Antigen-pulsed epidermal Langerhans cells protect susceptible mice from infection with intracellular parasite *Leishmania major*. Eur. J. Immunol. 28:3800.

50. Skeiky YA, Kennedy M, Kaufman D, Borges MM , Guderian JA, Scholler K, Ovendale PJ, Picha, KS, Morrissey PJ, Grabstein KH, Campos-Neto A and Reed SG 1999. LeIF: a recombinant Leishmania protein that induces an IL-12-mediated Th1 cytokine profile. J. Immunol. 161:61716179.

51. Padigel UM, Perrin P and Farrell JP. 2001. The development of a Th1 Type response and resistance to Leishmania major infection in the absence of CD40-CD40L costimulation. J Immunol.167: 5874-5879.

52. Muller I, Kropf P, Louis JA, and Milon G.1994. Expansion of gamma interferon-producing CD8+ T cells following secondary infection of mice immune to *Leishmania major*. Infect Immun. 62:2575-2581.

53. Farrell JP, Muller I and Louis JA.. 1989. A role for Lyt2+ T cells in resistance to cutaneous leishmaniasis in immunized mice. J. Immunol. 142:2052-2056,

54. Gurunathan S, Stobie L, Prussin C, Sacks DL, Glaichenhaus N, Iwasaki A, Fowell DJ, Locksley RM, Chang JT, Wu CY, and Seder RA. 2000. Requirements for the maintenance of Th1 immunity in vivo following DNA vaccination: a potential immunoregulatory role for CD8+ T cells. J Immunol. 165:915-924.

55. Overath P and Harbecke D. 1993. Course of Leishmania infection in beta 2-microglobulin-deficient mice. Immunol Lett. 37:13-17.

56. Huber M, Timms E, Mak TW, Rollinghoff M and Lohoff M. 1998. Effective and long-lasting immunity against the parasite *Leishmania major* in CD8-deficient mice. Infect Immun. 66:3968-3970.

57. Nabors, GS and Farrell JP. 1994. Depletion of interleukin-4 in BALB/c mice with established *Leishmania major* infections increases the efficacy of antimony therapy and promotes Th-1 like responses. Infection and Immunity 62: 5498-5504.

58. Nabors,GS, Alfonso LCC, Farrell J.P and Scott P. 1994. Switch from a type 2 to a type1 T helper cell response and cure of established *Leishmania major* infection in mice is induced by combined therapy with interleukin 12 and pentostam. P.N.A.S. USA 92:3142-3146.

59. Li J, Sutterwala S and Farrell JP. 1997. Successful therapy of chronic, nonhealing murine cutaneous leishmaniasis with sodium stibogluconate and gamma interferon depends on continued interleukin-12 production. Infection and Immunity 65:3225-3230.

60. Nabors GS and Farrell JP. 1996. Successful chemotherapy in experimental *Leishmania*sis is influenced by the polarity of the T cell response before treatment. J. Inf. Dis. 173:979-986.

61. Li J, Hunter CA and Farrell JP. 1999 Anti-TGF-ß treatment promotes rapid healing of *Leishmania major* infection in mice by enhancing in vivo nitric oxide production. J. Immunol.162:974-979.

63. Handman E, Noormohammadi AH, Curtis JM, Baldwin T, and Sjolander A. 2000. Therapy of murine cutaneous leishmaniasis by DNA vaccination. Vaccine 18. 3011-3017.

63. Kane MM, Mosser DM. 2001. The role of IL-10 in promoting disease progression in leishmaniasis. J Immunol. 166:1141-1147.

64. Belkaid Y, Hoffmann KF, Mendez S, Kamhawi S, Udey MC, Wynn TA, and Sacks DL. 2001. The Role of interleukin (IL)-10 in the persistence of *Leishmania major* in the skin after healing and the therapeutic potential of anti-IL-10 receptor antibody for sterile cure J Exp Med. 194:1497-1506.

65. DeKrey GK, Lima HC, and Titus RG. 1998 Analysis of the immune responses of mice to infection with *Leishmania braziliensis*. Infect Immun. 66:827-829.

66. Guevara-Mendoza O, Une C, Franceschi Carreira P, and Orn A. 1997. Experimental infection of Balb/c mice with Leishmania panamensis and Leishmania mexicana: induction of early IFN-gamma but not IL-4 is associated with the development of cutaneous lesions. Scand J Immunol. 46:35-40.

67. Almeida RP, Barral-Netto M, De Jesus AM, De Freitas LA, Carvalho EM, and Barral A. 1996. Biological behavior of *Leishmania amazonensis* isolated from humans with cutaneous, mucosal, or visceral leishmaniasis in BALB/C mice. Am J Trop Med Hyg. 54:178-184.

68. Afonso LCand Scott P. 1993. Immune responses associated with susceptibility of C57BL/10 mice to *Leishmania amazonens.*. Infect Immun. 61: 2952-2959.

69. Jones DE, Buxbaum LU, and Scott P. 2000. IL-4-independent inhibition of IL-12 responsiveness during *Leishmania amazonensis* infection. J Immunol. 2000;165:364-372

70. Soong L, Chang CH, Sun J, Longley BJ Jr, Ruddle NH, Flavell RA, and McMahon-Pratt D. 1997. Role of CD4+ T cells in pathogenesis associated with *Leishmania amazonensis* infection. J Immunol. 158:5374-5383.

71. Kima PE, Constant SL, Hannum L, Colmenares M, Lee KS, Haberman AM, Shlomchik MJ, and McMahon-Pratt D. 2000. Internalization of *Leishmania mexicana* complex amastigotes via the Fc receptor is required to sustain infection in murine cutaneous leishmaniasis. J Exp Med. 191:1063-1068.

72. Dedet JP, Pratlong F, Lanotte G, and Ravel C. 1999. Cutaneous leishmaniasis. The parasite. Clin Dermatol. 17:261-268.

73. Carvalho EM, Correia Filho D, Bacellar O, Almeida RP, Lessa H, and Rocha H. 1995. Characterization of the immune response in subjects with self-healing cutaneous leishmaniasis. Am J Trop Med Hyg. 53:273-277.

74. Rocha PN, Almeida RP, Bacellar O, de Jesus AR, Filho DC, Filho AC, Barral A, Coffman RL, and Carvalho EM. 1999. Down-regulation of Th1 type of response in early human American cutaneous leishmaniasis. J Infect Dis. 180:1731-1734.

75. Bourreau E, Prevot G, Pradinaud R, and Launois P. 2001. Interleukin (IL)-13 is the predominant Th2 cytokine in localized cutaneous leishmaniasis lesions and renders specific CD4+ T cells unresponsive to IL-12. J Infect Dis. 183:953-959.

76. Ajdary S, Alimohammadian MH, Eslami MB, Kemp K, and Kharazmi A. 2000. Comparison of the immune profile of nonhealing cutaneous leishmaniasis patients with those with active lesions and those who have recovered from infection. Infect Immun. 68:1760-1764.

77. Da-Cruz AM, Conceicao-Silva F, Bertho AL, and Coutinho SG. 1994. Leishmania-reactive CD4+ and CD8+ T cells associated with cure of human cutaneous leishmaniasis. Infect Immun. 62:2614-2618.

78. Toledo VP, Mayrink W, Gollob KJ, Oliveira MA, Costa CA, Genaro O, Pinto JA, and Afonso LC. 2001 Immunochemotherapy in American cutaneous leishmaniasis: immunological aspects before and after treatment. Mem Inst Oswaldo Cruz. 96:89-98.

79. Bottrel RL, Dutra WO, Martins FA, Gontijo B, Carvalho E, Barral-Netto M, Barral A, Almeida RP, Mayrink W, Locksley R, and Gollob KJ. 2001 Flow cytometric determination of cellular sources and frequencies of key cytokine-producing lymphocytes directed against recombinant LACK and soluble *Leishmania* antigen in human cutaneous leishmaniasis. Infect Immun.69:3232-3239.

80. Vouldoukis I, Riveros-Moreno V, Dugas B, Ouaaz F, Becherel P, Debre P, Moncada S, and Mossalayi MD. 1995. The killing of *Leishmania major* by human macrophages is mediated by nitric oxide induced after ligation of the Fc epsilon RII/CD23 surface antigen. Proc Natl Acad Sci U S A 92:7804-7808.

81. Mossalayi MD, Arock M, Mazier D, Vincendeau P, and Vouldoukis I. 1999. The human immune response during cutaneous leishmaniasis: NO problem. Parasitol Today. 15:342-345.

82. Convit J. 1996. Leishmaniasis: Immunological and clinical aspects and vaccines in Venezuela. Clin Dermatol. 14:479-787.

83. Weigle K and Saravia NG. 1996. Natural history, clinical evolution, and the host-parasite interaction in New World cutaneous Leishmaniasis. Clin Dermatol. 14:433-450.

84. Caceres-Dittmar G, Tapia FJ, Sanchez MA, Yamamura M, Uyemura K, Modlin RL, Bloom BR and Convit J. 1993. Determination of the cytokine profile in American cutaneous leishmaniasis using the polymerase chain reaction. Clin Exp Immunol. 91:500-505.

85. Pirmez C, Yamamura M, Uyemura K, Paes-Oliveira M, Conceicao-Silva F, and Modlin RL. 1993. Cytokine patterns in the pathogenesis of human leishmaniasis. J Clin Invest. 91:1390-1395.

86. Akuffo H, Maasho K, Blostedt M, Hojeberg B, Britton S, and Bakhiet M. 1997. *Leishmania aethiopica* derived from diffuse leishmaniasis patients preferentially induce mRNA for interleukin-10 while those from localized leishmaniasis patients induce interferon-gamma. J Infect Dis. 175:737-741.

87. Rodriguez V, Centeno M, and Ulrich M 1996. The IgG isotypes of specific antibodies in patients with American cutaneous leishmaniasis; relationship to the cell-mediated immune response. Parasite Immunol.18:341-345.

88. Bomfim G, Nascimento C, Costa J, Carvalho EM, Barral-Netto M, and Barral A. 1996. Variation of cytokine patterns related to therapeutic response in diffuse cutaneous leishmaniasis. Exp Parasitol. 84:188-194.

89. Barral A, Teixeira M, Reis P, Vinhas V, Costa J, Lessa H, Bittencourt AL, Reed S, Carvalho EM, and Barral-Netto M. 1995. Transforming growth factor-beta in human cutaneous leishmaniasis. Am J Pathol. 147:947-954.

90. Melby PC, Andrade-Narvaez FJ, Darnell BJ, Valencia-Pacheco G, Tryon VV, and Palomo-Cetina A. 1994. Increased expression of proinflammatory cytokines in chronic lesions of human cutaneous leishmaniasis. Infect Immun. 62:837-842.

91. Carvalho EM, Johnson WD, Barreto E, Marsden PD, Costa JL, Reed S, and Rocha H. 1985. Cell mediated immunity in American cutaneous and mucosal leishmaniasis. J Immunol. 135:4144-4148.

92. Cabrera M, Shaw MA, Sharples C, Williams H, Castes M, Convit J, and Blackwell JM. 1995. Polymorphism in tumor necrosis factor genes associated with mucocutaneous leishmaniasis. J Exp Med. 182:1259-1264.
93. Blackwell JM. 1999. Tumour necrosis factor alpha and mucocutaneous leishmaniasis. Parasitol Today 15:73-75.

ANTI-LEISHMANIA VACCINE

Antonio Campos-Neto

Infectious Disease Research Institute, Seattle, WA 98104, USA

INTRODUCTION

Vaccination against cutaneous leishmaniasis has been used or tested in humans for approximately 75 years. In the Old World, deliberated inoculation of virulent organisms from the pus of an active lesion (probably *L. major*) in non-exposed areas of the body is an ancient practice. Promastigotes of *L. major* grown in culture was first used in Russia in 1937 by Lawrow and Dubowokoj as a means to effectively induce protection against natural infection (1). More recently, standardized inoculum of culture promastigotes were developed by Israeli scientists and used in several trials (2). This process known as leishmanization is still used in some countries, notably Uzbekistan (3). Leishmanization has been proven to be efficacious against Old World cutaneous leishmaniasis (4). However, several basic and logistic problems have precluded the widespread use of this procedure to prevent cutaneous leishmaniasis. Some of these problems include: 1. Difficulty in standardizing the virulence of the vaccine; and 2. Side effects such as the severe and long lasting lesions that occur in many vaccinated individuals (5). Moreover, there are no evidences of the effectiveness of leishmanization against either New World tegumental leishmaniasis or against human visceral leishmaniasis. Vaccination using crude antigen preparation obtained from promastigote forms of various species of *Leishmania* have been tested in human clinical trials in both the Old and New World. The results vary from 0-75% efficacy against cutaneous leishmaniasis and only modest protection against visceral leishmaniasis (6-10). However, it is difficult to compare the results of these trials because they were performed in different areas of the World using different species of *Leishmania* and different adjuvants or no adjuvant at all.

Notwithstanding the logistic restrictions of leishmanization and the inconsistency of the trials performed with different crude antigenic preparation of leishmanial promastigote antigens one important conclusion though can be drawn from these studies: deliberate induction of protection against leishmaniasis is feasible and can be achieved with either a viable vaccine or with leishmanial sub-unit components.

In this chapter, we will emphasize new approaches and progress towards the development of a sub-unit vaccine against leishmaniasis with particular emphasis on purified leishmanial recombinant proteins.

NEW APPROACHES OF VACCINE DEVELOPMENT AGAINST LEISHMANIASIS
Recombinant leishmanial proteins
The promising results of some human clinical trials using different preparations and species of killed leishmania lend clear support for the development of a more defined anti-leishmania sub-unit vaccine. Several investigators, including us, have over the past decade, search for genes encoding leishmanial proteins that could induce protection against cutaneous and visceral leishmaniasis in several experimental models of the diseases. A great number of recombinant proteins have been described and obtained using a variety of lucidly planed cloning strategies. However, soon after the great success in producing these recombinant molecules it became evident that a critical step to successfully delivery them to induce the right type of immune response against leishmaniasis was missing, i.e., the induction of a strong antigen specific Th1 response in the absence of Th2 response. Most of the adjuvants used in these initial experiments, included both adjuvants licensed for human use such as alum, BCG, etc., and adjuvants used in animal experimentation. Unfortunately these adjuvants stimulate primarily Th2 type of responses. A variety of different strategies have been used aiming to circumvent this impasse. Some of them will be discussed in the next section.

Live vectors
Several types of host vectors have been described as carries for experimental vaccines. The general principle behind this means of antigen delivery, particularly for viral vectors, is fundamentally the stimulation of both CD4+ and CD8+ subsets of T cells. In contrast, most known adjuvants activate primarily CD4+ T cells. In addition, because all these vectors cause sub clinical infections, the immune responses that they induce are stronger than that induced by immunization with killed organisms. Moreover, they induce the same type of long last immunity that usually follows a cured infectious disease. Most of the vectors thus far tested are either virions or viable attenuated bacteria, which in general are vaccines themselves. Among the virus vectors the most common ones thus far tested are the *Vaccinia* virus and *Adenovirus* (11-15). Other virus like the lentivirus are promising candidates but there is still several problems with them that need to be solved before these vectors become available for both more ample research and for clinical trials (16). Among the bacterial vectors, the Bacille

Calmette-Guerin, an attenuated strain of *Mycobacterium bovis* (17, 18) and an auxotrophic mutant of *Salmonella typhi* (19-21) have been used by various groups as promising carriers of genes of a variety of infectious organisms including *Leishmania*. However, the delivery of antigens with live vectors has in general some serious disadvantages. For example BCG, despite its built in adjuvanticity and its overwhelmingly use in humans for more than 75 years, it can cause severe diseases in immunocompromised persons(22-26); *Vaccinia* virus, despite its proven excellence in inducing protection against smallpox; the long term immunological memory that the virus induces in both humans and experimental animals; the appropriateness of its genome to carry several different foreign genes, the virus can cause rare but severe side effects that limits its widespread use in the population, particularly as a vehicle of diseases with morbidity smaller than the side effects of the vaccine. The adenovirus, despite its overall lack of virulence for humans; of a genome that can accept several foreign genes; and the great attractiveness of the oral route of vaccine delivery, most humans have had sub clinical infections with this virus (27, 28). As a consequence most humans have high antibody titers against the virus, which interferes with the colonization of these organisms and consequently with the effectiveness of the vaccine (29, 30).

To date, BCG, *Vaccinia* virus, and *Salmonella typhi*, have been tested as leishmanial antigens delivering vectors. BCG is an attractive system in that this *Mycobacterium* has already been used as adjuvant in human trials against leishmaniasis using killed leishmania as antigen (9, 10, 31-33). Particularly for the immunotherapy of cutaneous leishmaniasis, addition of BCG to killed leishmania vaccine resulted in faster healing than treatment with killed leishmania alone (31). Moreover, PBMC from individuals of the former group, upon stimulation in vitro with parasite's antigens, produced high levels of IL-2 and IFN-γ, thus suggesting a Th1 type of response. Live recombinant BCG expressing the leishmania surface gp63 protein has been engineered and tested for protection in the murine model of cutaneous leishmaniasis. In mice of the resistant phenotype (CBA/J) the BCG construct induced substantial protection against challenge with either *L. major* or *L. mexicana*. However, in susceptible mice (BALB/c), vaccination with BCG expressing gp63 while inducing protection against challenge with *L. mexicana*, it resulted in only a delay of the infection upon challenge with *L. major* (34, 35). More recently recombinant BCG expressing the *L. chagasi* protein LCR1 was evaluated as vaccine candidate and shown to induce only partial protection in the murine model of visceral leishmaniasis. Moreover, the protection was dependent of the route of the vaccine delivery (36).

The *Vaccinia* vector has been tested as a delivery system of a variety of antigens. One of them included the *L. amazonensis* gp46 gene. This gene was cloned into an attenuated mutant of the *Vaccinia* virus and used in protection experiments in susceptible BALB/c mice against challenge with *L. amazonensis*.

Long-term memory (up to 10 months post immunization) was observed in these animals. Unfortunately, significant levels of protection were observed only when the mice were challenged with a low dose (10^3) of parasites (37).

The *S. typhi* auxotrophic mutants similarly to the recombinant BCG constructs have been tested as carrier vectors for the leishmanial gp63 gene. One interesting attraction of the salmonella system is the oral route of vaccine delivery. The mutants are highly attenuated because they are dependent of exogenous aromatic amino acids, but they can still colonize and penetrate the intestinal mucosal surface and reach the local lymphoid structures of the gastro-intestinal tract. Vaccinated mice develop strong and protective antibody and T cell responses against challenge with highly virulent strains of *Salmonella*. The leishmanial gp63 gene was cloned in these vectors and tested as vaccine candidate against challenge with *L. major* in both resistant and susceptible mice strains. Substantial protection was achieved in resistant mice. However, in BALB/c mice, the results are conflicting. In one report (38), which used the auxotrophic mutant AroA⁻ (deletion of the 5-enolpyruvylshikimate 3-phosphate synthase gene) no protection was achieved. In contrast, in other reports (39, 40), protection was achieved using gp63 expressed in a double auxotrophic *S. typhi* mutant containing the *AroA* and *AroD* (3-dehydroquinase gene) gene deletions. There is no clear explanation to understand these discordant results. More recently (41), the *AroA⁻AroD⁻* double mutant expressing gp63 of *L. mexicana mexicana* was tested in F1 (BALB/cXC57BL/6) mice (resistant phenotype) and induced significant protection against the challenge with a wild type strain of *L. mexicana mexicana*.

In contrast to the general optimism generated by the initial results using both bacterial and viral vectors as antigen delivery systems for an anti-leishmanial vaccine, the outcome of the subsequent experiments was rather disappointing. Regardless of the vector used no substantial protection could be consistently be induced in mice of the susceptible phenotype (BALB/c) challenged with *L. major*. In all cases, the inefficiency of the vaccination could not be attributed to the generation of an inadequate immune response modulated by the infecting vectors. Strong and specific Th1 responses were stimulated by the vaccination with BCG carrying the leishmanial antigens gp63 or LCR1, with *Vaccinia* virus carrying gp46, and *S. typhi* carrying gp63 and yet no protection was observed (34, 36-38). It is interesting to mention that none of these three antigens, even in combination with IL-12 induce protection in BALB/c mice challenged with *L. major*. IL-12 has been considered the gold standard Th1 adjuvant and it has been successfully utilized in protection experiments against leishmaniasis with both crude and recombinant leishmanial antigens in both the murine and non-human primate models of human diseases (42-44). Thus it is possible that the inefficiency of this process of antigen delivery is not related to the system itself but rather because very few and non-protective antigens were tested. This possibility can be easily tested with protective recombinant antigens such as LmSTI1, TSA, and LACK.

Naked plasmid DNA

Over the past decade immunization with plasmid DNA containing genes from a variety of infectious agents gained enormous popularity particularly because this means of antigen delivery is very efficient to stimulate CD8+ T cell response specific for the proteins encoded by such genes. In addition, antibody and CD4+ T cell responses preferentially of the Th1 phenotype are generated by DNA immunization. The biological and molecular mechanisms involved in the interiorization of the bacterial plasmid DNA followed by the subsequent transcription and translation for the protein production and expressed by the mammalian cells are unclear. However, the plasmid DNA is readily taken up by antigen presenting cells that efficiently produce and process the plasmid encoded protein in both the Class I and Class II pathways of the antigen processing machinery (45, 46).

The success of DNA immunization is dependent of the route of administration and is largely a function of the expression of the foreign antigen by the transduced APC. Thus it is well defined that sub-cutaneous and intra-muscular administration of plasmid DNA result in the induction of CTL, Ab, and CD4 response primarily of the Th1 phenotype. In contrast, intra-dermal immunization, particularly with a gene gun results in CTL, Ab, and CD4 response primarily of the Th2 phenotype (47-50). This dichotomy to Th1 or Th2 responses are not entirely understood but it has been suggested that it is related to the different APCs that presents and initiates the immune response in the dermal versus in the subcutaneous and muscles tissues (51).

Expression of the cloned foreign gene present in the plasmid is unquestionable an important factor for the immunogenicity of the DNA immunization. We have tested DNA immunization with several leishmanial genes and observed that immunogenicity and protection against challenge of mice with *L. major* are chiefly related with the levels of the leishmanial protein that is expressed in vitro by an eukaryotic cell (APC) transfected with the plasmid containing the referred gene. For example, pcDNA3 containing the leishmanial genes TSA, LmSTI1, Ldp23 and LeIF genes induce different levels of protection in BALB/c mice challenged with *L. major*. Thus, TSA-pcDNA induces excellent protection, LmSTI1-pcDNA induces moderate protection and both Ldp23-DNA and LeIF-pcDNA induce no protection. Coincidentally, the in vitro expression of the recombinant proteins by transfected macrophages is as follows: TSA is abundantly expressed in transfected cells (detected by both Coomassie stained PAGE or by western); LmSTI1 is expressed in amounts not detectable by Coomassie stained PAGE but is easily detected by western blot analysis; Ldp23 and LeIF are hardly or not at all detected in transfected macrophages (unpublished observations). To date, it is not clear why APCs transfected with a plasmid DNA encoding different genes under the control of the same promoter (for example CMV for the above experiments) end up expressing highly different concentrations of the encoding proteins. For more information on these and on several other basic aspects of DNA immunization the reader should refer a series of comprehensive review articles (45, 52-54).

Notwithstanding the mechanisms by which bacterial plasmid DNA immunization results in both, the expression of antigens in mammalian eukaryotic cells and in selective induction of various arms of the immune system, this means of antigen delivery has been successfully proven as promising alternative in vaccine development against a variety of infectious diseases. For example, DNA immunization has been reported to have potential as vaccine candidates against numerous virus including influenza, rabies, HIV, HBC, HCV, and HSV (55-60). In addition, several bacterial and protozoa vaccine candidates have also tested including *Mycobacterium tuberculosis* (61-63), *Mycoplasma pulmonis* (64), *Plasmodium* (65, 66), *Leishmania* (67-73), and *Trypanosoma cruzi* (74-76).

DNA immunization against leishmaniasis is perhaps one of the most apparent examples of the efficiency of this method of antigen delivery. Several genes/antigens have been tested and shown to induce excellent protection in highly susceptible BALB/c mice challenged with *L. major* (67-73). Moreover several of these studies have shown that DNA immunization can achieve sterilizing protection (67). In these studies, no parasite could be cultured from the site of challenge and protected mice did not serve as subclinical reservoirs of the disease because when tested using sand flies, the natural vectors of the diseases, transmission could not be obtained (67). In addition, DNA immunization seems to confer very long-term protection (77, 78).

These results open exciting possibilities for the future use of DNA vaccine against human leishmaniasis. Several studies are under investigation to expand the results from the murine model to dogs and non-human primates models of these diseases. Thus far the results obtained from the non-human primate studies have not reproduced the high level of protection obtained in mice with plasmid DNAs like TSA-DNA and LACK-DNA (unpublished observations). No results on the dog model are available at this point. However, despite these initial disappoint results several groups are encouraged to continue these studies vis-à-vis the great success of this means of immunization in protection experiments carried out in monkeys against SIV (79). It seems that successful DNA immunization in non-human primates, and for that matter in humans as well, need codon optimization of the DNA for these species (80). In the latter experiments, successful protection against SIV was only achieved after optimized DNA was used to vaccinate the monkeys. Therefore, it is fair to assume that the results of the experiments thus far conducted in monkeys with leishmanial DNA are too preliminary to merit conclusive interpretation.

Immunomodulatory adjuvants

Before 1995 basically all adjuvants thus described were aseptic tissue irritants intended to stimulate a strong productive inflammatory reaction at their injection site i.e., a depot effect that created the optimal conditions to augment the overall immune response to the antigen. However, after the definition of the opposite effects that Th1 and Th2 immune responses can exert in the outcome of several different types of infections and particularly in leishmaniasis, it became evident the need of adjuvants that could preferentially induce either Th1 or Th2 immune responses.

IL-12: The idea of an immunomodulatory adjuvant was first turn into reality when Scott et al. (42) elegantly showed that the cytokine IL-12 was a powerful physiological adjuvant for the selective stimulation of Th1 responses if administered mixed with the antigen. IL-12 injected sub-cutaneously with leishmanial soluble antigens (SLA), induced strong anti-SLA Th1 response and no detectable Th2 response to this antigen. Importantly, this protocol of immunization conferred excellent protection in BALB/c mice challenged with *L. major*. IL-12 triggers the Th1 responses because it binds to its receptors on the cell surface of the uncommitted Th0 CD4+ T cells and stimulates them, to differentiate into IFN-γ producing cells (81-85). Therefore, the presence of IL-12 in the milieu of the antigen stimulation of the T cells by the APCs, provide the necessary signal that causes the responding antigen specific T cells to differentiate to Th1 phenotype (82). IL-12 has been successfully used as Th1 adjuvant for a variety of antigens in both the murine and in the non-human primates models of several infectious diseases including leishmaniasis (42-44, 86-88).

One of the drawbacks of IL-12 is the fact that its adjuvant effect is practically restricted to the modulation of the immune response to the Th1 phenotype i.e., there is no depot effect and consequently the amplitude of the immune response is not augmented as it is for conventional adjuvants. Moreover, in contrast with conventional adjuvants and to DNA immunization, it seems that the immunological memory to the immunizing antigen is not stimulated appropriately when IL-12 is used as adjuvant. Thus, vaccination of BALB/c mice with the leishmanial antigen LACK mixed with IL-12 as adjuvant resulted in short term protection against challenge with *L. major*. In contrast vaccination with LACK-DNA induced long-term protection. In addition, these studies suggested that the short-term protection could not be overcome by creating a depot condition after the mice immunization with LACK plus IL-12. Thus mice immunized with a mixture of LACK + IL12 + Alum (added as a component of the vaccine to create the depot effect) developed a strong Th1 response to LACK but the short-term protection could not be overcome by the generation of depot (73, 77). These results question the utility of IL-12 for general use in vaccine development. Moreover, at the present time the manufacture of IL-12 in large scale is difficult and expensive imposing serious logistics restrictions to its use in mass vaccination in humans.

Oligonucleotides containing CpG motifs: Soon after plasmid DNA immunization wa discovered as an alternative form of antigen delivery, several investigators observed that many preparations of these molecules acted as non-specific activators of various arms of the innate immunity. Thus it was observed that plasmid DNA could act as polyclonal activators of B cells, stimulate up-regulation of co-stimulatory molecules, and activate macrophages for the production of IL-12, inflammatory cytokine, as well as the production of oxidant radicals such as the nitric oxide (NO). It is now clear that these immunostimulatory properties of bacterial DNA can be recapitulated by synthetic oligodeoxinucleotides containing specific motifs centered on CpG dinucleotide (CpG ODN) sequences (89-93). The molecular mechanisms of the immunostimulation caused by these molecules have not yet been elucidated. However, pre-clinical studies have shown that CpG ODN can enhance innate immunity against a variety of infectious organisms and act as immunomodulatory adjuvants as well. Indeed, several recent studies have shown that injection of BALB/c mice with CpG ODN without antigens induces a state of partial resistance in these animals up to five weeks against challenge with *L. major* (94). If the CpG ODN is injected in conjunction with leishmanial soluble antigen (SLA) significant protection is obtained in these animals that is maintained for as long as six months. In these experiments, the immunostimulatory properties of the CpG ODN were associated with the emergence of strong Th1 response to SLA (95, 96). In contrast mice that were immunized with SLA alone developed a Th2 response and when challenged with *L. major* developed an aggravated disease when compared with control non-immunized animals. Moreover, it was observed that this in vivo action of the CpG ODN is dependent of a systemic stimulation of IL-12 production, which ultimately leads to the generation of the Th1 response to SLA. Interestingly CpG ODN had also a substantial therapeutic effect in BALB/c mice infected with *L. major* (96). Similarly to its adjuvant activity in the prophylaxis experiments, the curative effect of CpG ODN was dependent of systemic induction of IL-12. Thus, administration of neutralizing anti-IL-12 mAb (or mAb anti-IFN- γ for that matter) before inoculation of the oligonucleotydes totally abrogated its curative effect. Therefore, because of CpG ODN have both prophylactic and therapeutic properties these molecules are of considerable interest in vaccine development for both prevention and treatment of diseases like leishmaniasis.

Unfortunately, thus far the immunostimulatory actions of the CpG ODN, similarly to plasmid DNA, have not been reproduced successfully beyond the mouse model. Several attempts have been made by many investigators to optimize or to find the right CpG sequences that are as effective in humans as they are in mice. Some promising pre-clinical findings have been reported (97-99). However, it is still premature to evaluate if these results can be translated into effective and safe intervention.

Conventional adjuvants: Over the past few years, a great number of conventional adjuvant formulations that stimulate preferentially some arms of the immune system have been described. Examples of such adjuvants include bacterial cell wall components like the muramyl peptide N-acethylmuramyl-L-alanyl-D-isoglutamine (MDP) and the monophosporyl lipid A (MPL®), saponins isolated from the bark of *Quillaja saponaria* (Quil A and QS-21), and mannide monooleate oil emulsions (Montanide® ISA 720) etc. Most of these products are stable and biodegradable water-in-oil emulsions. The mechanism of action of these formulations is totally obscure at the present time. However these adjuvants are highly attractive in that they are in general not toxic (or of very low toxicity) and are suitable for human use. A detailed and extensive description of these and other potentially human adjuvants can be found elsewhere (100, 101). We have systematically tested one of such adjuvants (MPL®) in conjunction with several leishmanial recombinant antigens in several animal models. The results are very encouraging and a human clinical trial with this adjuvant and the antigens TSA, LmSTI1, and LeIF is in progress (see below).

OTHER ALTERNATIVES
Auxotrophic mutants

Live attenuated vaccines are perhaps the oldest and most successful means of immunoprophylaxis against several infectious diseases (e.g. small pox, poliomyelitis, yellow fever, etc). Leishmanization, while not a procedure that uses an attenuated vaccine, is nonetheless a clear example that live organisms is a viable alternative for vaccine development against leishmaniasis. Several approaches have been used over the past century to obtain infective but non-virulent organisms for the development of attenuated vaccines. Induction and selection of naturally occurring mutants has been the central developmental process to generate these vaccines. However, these are empirical processes and only rarely they originate a strain of organism suitable to be used as safe and efficacious vaccine. In most cases the selected organism is not attenuated enough and consequently causes disease. Conversely, the attenuation is frequently so profound that it prevents implantation or colonization of the microbe in the host resulting in lack of adequate immunization.

Recent developments have made possible, through genetic intervention, the construction of stable auxotrophic mutants that can be an attractive approach to generate attenuated vaccine candidates for several infectious diseases including leishmaniasis. Indeed one such leishmania auxotrophic mutant has been generated and shown to protect BALB/c mice from infection with *L. major* (102). In this model, the *L. major* gene dihydrofolate reductase-thymidilate synthase (DHFR-TS) was deleted by target gene replacement (102) resulting in an organism auxotrophic for thymidine. This mutant maintained its ability to infect macrophages (possibly using thymidine from these cells) and differentiate into amastigotes both in vitro and in vivo. Interestingly, the DHFR-TS mutant lost its virulence but survived in BALB/c mice for up to two months, declining with a half-life of 2-3 days. More importantly, this mutant induced substantial resistance in both genetically susceptible and resistant mice challenged with virulent *L. major* (103) and partial protection in mice challenged with *L. amazonensis*. However, vaccination of rhesus monkeys with the DHFR-TS auxotrophic *L. major* resulted in no protection against challenge of these animals with virulent *L. major* (Grimaldi, G. Jr. et al, personal communication). These results yet disappointing do not invalidate the gene target approach for the development of auxotrophic attenuated mutants. Gene knockout of alternative molecule candidates that have lesser impact in the establishment of sub-clinical infections by the parasites can in principle result in a more efficacious and safe live attenuated leishmania vaccine candidate.

Sand Fly Saliva

Recent observations point to the sand fly saliva as an intriguing and interesting target for vaccine development against leishmaniasis. Using the mouse model, Kamhawi S. et al., have recently shown a protective effect of the saliva from the sand fly *Phebotomus papatasi* against challenge of both C57BL/6 and BALB/c mice with *L. major* (104-106). The mechanism of this protection has not yet been completely elucidated. However, it has been suggested that immunization of mice with *P. papatasi*'s saliva induces a strong delayed type hypersensitivity to the saliva components. The activated macrophages present in the milieu of the local immune mediated inflammatory response apparently kills the infective promastigotes injected with a subsequent blood meal of an infected sand fly (107). A gene encoding a protein of *P. papatasi*'s saliva has been identified and cloned. This molecule delivered in a naked DNA format induces significant protection against infection of C57BL/6 mice with *L. major* (108). This finding can have important repercussion in vaccine development not only against leishmaniasis but also against other vector borne diseases like malaria and Chagas' disease. However, further studies are needed to validate this approach.

PROGRESS TOWARDS DEVELOPMENT OF AN ANTI-LEISHMANIA SUB-UNIT VACCINE CONSISTED OF LEISHMANIAL RECOMBINANT PROTEINS

Over the past decade many *Leishmania* genes have been cloned and expressed as recombinant proteins. Most of these proteins have been tested as vaccine candidates in mice against cutaneous and visceral leishmaniasis using a variety of adjuvants including *Corynaebacterium parvum*, Detox, 4'-monophosphoryl lipid A, QS-21, *Mycobacterium bovis*-BCG, and IL-12 (109-113). With the exception of a few studies (110, 114) the results were in general disappointing in that protection was achieved only in mice of the resistant phenotype. When these antigens were tested in BALB/c mice against challenge with *L. major*, only partial or no protection was observed. Unfortunately there are no follow up reports on the efficacy or utility of the recombinant antigens that induce good protection in susceptible mice challenged with *L. major*.

It is important to point out that despite the fact that most of these anti-leishmanial vaccine candidate antigens have been discovered and selected based on the Th1 paradigm they turn out to be non-protective antigens. For example, BALB/c mice infected with *L. major* develop low antibody titers to the antigen Ldp23, and to many others that we have tested. In addition, Ldp23 induces preferentially Th1 response in lymph node cells from these infected mice (115). However, in combination with adjuvants that preferentially induce Th1 responses such as IL-12, despite stimulating strong antigen specific Th1 response in the absence of any detectable Th2 response, Ldp23 did not confer protection (unpublished observations). In contrast, the antigen LACK stimulates strong and preferentially Th2 responses in lymph node and spleen cells of *L. major*-infected BALB/c mice (116). In addition the sera of these animals contain high titers of IgG1 anti-LACK antibodies. In spite of this, LACK induces substantial protection in BALB/c mice if administered in conjunction with adjuvants that stimulate Th1 responses such as IL-12 (78, 114). In another situation, antigens like LmSTI1 stimulate a mixed but preferentially Th1 responses in lymph node cells of BALB/c mice infected with *L. major* and the sera of these animals contain high titers of both IgG1 and IgG2a anti-LmSTI1 antibodies. Notwithstanding, LmSTI1 induces excellent protection in these animals if used with IL-12 as adjuvant (44).

Therefore, for antigen selection purposes in vaccine development against leishmaniasis these results do not support the Th1 paradigm. On the other hand, these results by no means challenge the concept that a Th1 response is essential for protection against leishmaniasis. Indeed, as mentioned above, shifting to Th1 the immune response to LACK before infection, results in protection. Conversely, immunization of BALB/c mice with LmSTI1 formulated with alum (an adjuvant that stimulates preferentially a Th2 response) instead of IL-12, results in no protection of these animals when they are challenged with *L. major* (unpublished observations).

In conclusion, for vaccine development against leishmaniasis the polarization to Th1/Th2 antigen-specific immune response that is developed against the parasite antigens during the infectious process seems irrelevant. Rather, immunogenicity and perhaps the amount of antigen expressed or secreted by the parasite in vivo are more important factors reflecting protective anti parasite immune response. Antigens that fulfill these criteria should therefore be more successful in inducing protection as long as they are administered with adjuvants that in combination with them modulated a strong Th1 response.

With this in mind we have systematically tested the protection potential a variety of leishmanial recombinant antigens, notwithstanding the phenotype of the immune response they stimulate on cells of either patients or mice infected with *Leishmania*. At least 20 different recombinant antigens were tested. Out of those, two (LmSTI1 and TSA) induced excellent protection in both BALB/c mice (susceptible strain) and in the non-human primate (rhesus monkeys) model of cutaneous leishmaniasis (44). These antigens will soon be tested in human clinical trials as vaccine candidates against human leishmaniasis.

Pre-clinical studies with LmSTI1 and TSA

As mentioned before, IL-12 has been generally used in animal experimentation as the adjuvant of choice for the induction of Th1 response. For our pre-clinical studies, BALB/c mice were immunized twice (3 weeks apart) sub-cutaneously, with the antigens LmSTI1 and TSA mixed with murine recombinant IL-12. Mice were infected with *L. major* and disease development was monitored for the next three months. Fig. 1 illustrates the results and clearly shows that mice immunized with either LmSTI1 or TSA mixed with IL-12 are protected against the infection. In contrast, mice immunized with the antigens mixed with saline develop severe lesions. LmSTI1 as a single antigen induces excellent protection and could, perhaps in itself constitute a vaccine against leishmaniasis. However, a cocktail composed of LmSTI1 and TSA should be a better vaccine because by mixing the antigens an amplification of the parasite epitopes involved in the induction of protective anti-leishmania immune response is achieved. This is a desirable condition because a vaccine containing a broad range of different protective epitopes is unlikely to suffer from MHC related unresponsiveness even in heterogeneous populations such as humans and dogs.

The efficacy of LmSTI1 and TSA in the non-human primate model of leishmaniasis was tested in *Macaca mulatta* (rhesus monkeys) This monkey model, though used to a much lesser extent than the mouse model, has been accepted as a system that more closely mirrors human immunity for vaccine development against several infectious diseases (43, 117-122).

Monkeys were vaccinated twice, one month apart, with a vaccine preparation containing LmSTI1 plus TSA mixed with the recombinant human IL-12 (Genetic Institute, Cambridge, MA), and 200_g of alum (Rehydragel HPA, Brekeley Heights, NJ) as adjuvant (43). These animals mounted excellent protection against challenge with 10^7 metacyclic promastigotes of *L. major* (Fig. 2).

Despite of the limitations of using IL-12 as adjuvant (very expensive product and difficult to manufacture), these studies are highly relevant for vaccine development against the human diseases because of the excellent protection that is achieved with purified recombinant antigens in two different animal models of the human disease. In this regard, it is important to emphasize that the protective properties of LmSTI1 and TSA antigens are not simply an ubiquitous phenomenon consequent to the

Figure 1.   *Vaccination of BALB/c mice against L. major infection with recombinant leishmanial proteins.*

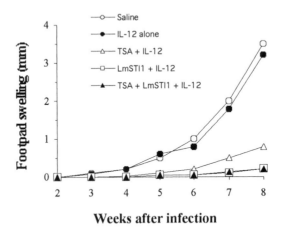

Weeks after infection

Mice (five per group) were immunized s.c., in the left footpad, twice (3 weeks apart) with 10μg of LmSTI1 or TSA mixed with 1μg of IL-12. Three weeks after the last immunization the mice were infected in the right footpad with 10^4 amastigote forms of L. major and footpad swelling was measured weekly thereafter.

modulation of the mouse immune response to the Th1 phenotype by the adjuvant IL-12. As mentioned before, we have tested this adjuvant with dozens of different leishmanial recombinant antigens and protection could only be observed when LmSTI1 and/or TSA were used as antigens. Therefore, in addition to inducing Th1 type of response, triggering of protective parasite epitopes per se is a crucial ingredient to elicit protection against leishmaniasis. These observations support our current view on the actual value of using the paradigm of Th1/Th2 immune response (see above) in vaccine development against leishmaniasis.

Clinical trials

Recent studies from our group using mice, dogs, and monkeys as animal models of the human diseases (unpublished) point to high immunogenicity and protective efficacy of the antigens LmSTI1 and TSA formulated with the adjuvant MPL-SE® (Corixa Corp., Seattle, WA). MPL® is a de-toxified 4'-monophosphoryl lipid A derivative of LPS obtained from *Salmonella minnesota*. Several studies have demonstrated that this product is a potent immunostimulant and yet lacks the toxic properties of LPS. In addition, immunization of mice with several different antigens formulated with MPL® results in a predominant Th1 response as determined by both in vivo and in vitro studies (123). MPL® has been tested in thousands of individuals, in several Phase I and Phase II clinical trials and proven to be a safe and efficacious adjuvant for human use.

Figure 2. *Protection of rhesus monkeys against cutaneous leishmaniasis by vaccination with the leishmanial recombinant antigens LmSTI1 and TSA formulated with recombinant human IL-12 and alum as adjuvants.*

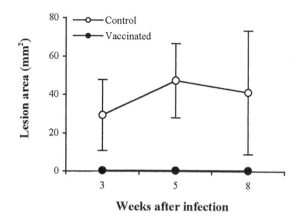

Monkeys were immunized 3x (approximately one month apart) with f LmSTI1 and TSA mixed with alum. Control monkeys were injected with saline only. Forty days after the last immunization, monkeys (six animals per group) were infected intra-dermally, in the left upper eyelid with 10^7 metacyclic promastigotes of *L. major*. Lesion sizes were measured at three weeks intervals and are expressed in mm^2 +/- SD.

In view of these results, Phase I/II clinical trials lead by Corixa Corporation in collaboration with the Infectious Disease Research Institute are in progress (presently Phase I) to test the efficacy of a vaccine consisted of the above recombinant antigens formulated with the adjuvant MPL-SE® The results of these trials should be available by the end of the year 2003.

REFERENCES

1.Modabber, F., M. L. Chance, D. A. Evans, E. E. Zijstra, A. M. El Hassan, P. A. Kager, R. W. Ashford, N. C. Hepburn, A. H. S. Omer, and J.-P. Dedet. 1999. The Leishmaniasis. In Protozoal Diseases. Herbert M.Gilles, ed. Arnold, London, Sydney, Auckland, pp. 413-529.
2.Handman, E., D. T. Spira, A. Zuckerman, and B. Montilio. 1974. Standardization and quality control of *Leishmania tropica* vaccine. J. Biol. Stand. 2:223.
3.Sergiev, V. P. 1992. Control and prophylaxis of cutaneous leishmaniasis in the middle Asia republics of the former USSR. Bull. Soc. Fran. Parasit. 10:183.

4.Nadim, A., E. Javadian, G. Tahvildar-Bidruni, and M. Ghorbani. 1983. Effectiveness of leishmanization in the control of cutaneous leishmaniasis. Bull Soc. Pathol. Exot. Filiales. 76:377.

5.Modabber, F. 1989. Experiences with vaccines against cutaneous leishmaniasis: of men and mice. Parasitology 98 Suppl:S 49-S60.

6.Convit, J. 1996. Leishmaniasis: Immunological and clinical aspects and vaccines in Venezuela. Clin. Dermatol. 14:479.

7.Antunes, C. M., W. Mayrink, P. A. Magalhaes, C. A. Costa, M. N. Melo, M. Dias, M. S. Michalick, P. Williams, A. O. Lima, J. B. Vieira, and . 1986. Controlled field trials of a vaccine against New World cutaneous leishmaniasis. Int. J. Epidemiol 15:572.

8.Armijos, R. X., Weigel, M. M., Aviles, H., Maldonado, R. & Racines, J. 1998. Field trial of a vaccine against New World cutaneous leishmaniasis in an at-risk child population: safety, immunogenicity, and efficacy during the first 12 months of follow-up. J. Infect. Dis 177: 1352-1357.

9.Sharifi, I., A. R. FeKri, M. R. Aflatonian, A. Khamesipour, A. Nadim, M. R. Mousavi, A. Z. Momeni, Y. Dowlati, T. Godal, F. Zicker, P. G. Smith, and F. Modabber. 1998. Randomised vaccine trial of single dose of killed *Leishmania major* plus BCG against anthroponotic cutaneous leishmaniasis in Bam, Iran. Lancet 351:1540.

10.Khalil, E. A., A. M. El Hassan, E. E. Zijlstra, M. M. Mukhtar, H. W. Ghalib, B. Musa, M. E. Ibrahim, A. A. Kamil, M. Elsheikh, A. Babiker, and F. Modabber. 2000. Autoclaved *Leishmania major* vaccine for prevention of visceral leishmaniasis: a randomised, double-blind, BCG-controlled trial in Sudan. Lancet 356:1565.

11.Randrianarison-Jewtoukoff, V. and M. Perricaudet. 1995. Recombinant adenoviruses as vaccines. Biologicals 23:145.

12.Zhang, W. W. 1999. Development and application of adenoviral vectors for gene therapy of cancer. Cancer Gene Ther. 6:113.

13. Babiuk, L. A. and S. K. Tikoo. 2000. Adenoviruses as vectors for delivering vaccines to mucosal surfaces. J. Biotechnol. 83:105.

14.Paoletti, E. 1996. Applications of pox virus vectors to vaccination: an update. Proc. Natl. Acad. Sci. U. S. A 93:11349.

15.Walther, W. and U. Stein. 2000. Viral vectors for gene transfer: a review of their use in the treatment of human diseases. Drugs 60:249.

16.Kafri, T. 2001. Lentivirus vectors: difficulties and hopes before clinical trials. Curr. Opin. Mol. Ther. 3:316.

17.Falk, L. A., K. L. Goldenthal, J. Esparza, M. T. Aguado, S. Osmanov, W. R. Ballou, S. Beddows, N. Bhamarapravati, G. Biberfeld, G. Ferrari, D. Hoft, M. Honda, A. Jackson, Y. Lu, G. Marchal, J. McKinney, and S. Yamazaki. 2000. Recombinant bacillus Calmette-Guerin as a potential vector for preventive HIV type 1 vaccines. AIDS Res. Hum. Retroviruses 16:91.

18.Gicquel, B. 1995. BCG as a vector for the construction of multivalent recombinant vaccines. Biologicals 23:113.

19.Fooks, A. R. 2000. Development of oral vaccines for human use. Curr. Opin. Mol. Ther. 2:80.

20.Levine, M. M., J. Galen, E. Barry, F. Noriega, S. Chatfield, M. Sztein, G. Dougan, and C. Tacket. 1996. Attenuated Salmonella as live oral vaccines against typhoid fever and as live vectors. J. Biotechnol. 44:193.

21.Hone, D. M., G. K. Lewis, M. Beier, A. Harris, T. McDaniels, and T. R. Fouts. 1994. Expression of human immunodeficiency virus antigens in an attenuated Salmonella typhi vector vaccine. Dev. Biol. Stand. 82:159.

22.Cunningham, J. A., J. D. Kellner, P. J. Bridge, C. L. Trevenen, D. R. Mcleod, and H. D. Davies. 2000. Disseminated bacille Calmette-Guerin infection in an infant with a novel deletion in the interferon-gamma receptor gene. Int. J. Tuberc. Lung Dis 4:791.

23.Abramowsky, C., B. Gonzalez, and R. U. Sorensen. 1993. Disseminated bacillus Calmette-Guerin infections in patients with primary immunodeficiencies. Am. J. Clin. Pathol. 100:52.

24.Jouanguy, E., F. Altare, S. Lamhamedi, P. Revy, J. F. Emile, M. Newport, M. Levin, S. Blanche, E. Seboun, A. Fischer, and J. L. Casanova. 1996. Interferon-gamma-receptor deficiency in an infant with fatal bacille Calmette-Guerin infection. N. Engl. J. Med. 335:1956.

25.Newport, M. J., C. M. Huxley, S. Huston, C. M. Hawrylowicz, B. A. Oostra, R. Williamson, and M. Levin. 1996. A mutation in the interferon-gamma-receptor gene and susceptibility to mycobacterial infection. N. Engl. J. Med. 335:1941.

26.Uysal, G., M. A. Guven, and O. Sanal. 1999. Subcutaneous nodules as presenting sign of disseminated BCG infection in a SCID patient. Infection 27:293.

27.Piedra, P. A., G. A. Poveda, B. Ramsey, K. McCoy, and P. W. Hiatt. 1998. Incidence and prevalence of neutralizing antibodies to the common adenoviruses in children with cystic fibrosis: implication for gene therapy with adenovirus vectors. Pediatrics 101:1013.

28.Schmidt, N. J., E. H. Lennette, and C. J. King. 1966. Neutralizing, hemagglutination-inhibiting and group complement-fixing antibody responses in human adenovirus infections. J. Immunol. 97:64.

29.Moffatt, S., J. Hays, H. HogenEsch, and S. K. Mittal. 2000. Circumvention of vector-specific neutralizing antibody response by alternating use of human and non-human adenoviruses: implications in gene therapy. Virology 272:159.

30.Xiang, Z. Q., Y. Yang, J. M. Wilson, and H. C. Ertl. 1996. A replication-defective human adenovirus recombinant serves as a highly efficacious vaccine carrier. Virology 219:220.

31.Convit, J., P. L. Castellanos, A. Rondon, M. E. Pinardi, M. Ulrich, M. Castes, B. Bloom, and L. Garcia. 1987. Immunotherapy versus chemotherapy in localised cutaneous leishmaniasis. Lancet 1:401.

32.Momeni, A. Z., T. Jalayer, M. Emamjomeh, A. Khamesipour, F. Zicker, R. L. Ghassemi, Y. Dowlati, I. Sharifi, M. Aminjavaheri, A. Shafiei, M. H. Alimohammadian, R. Hashemi-Fesharki, K. Nasseri, T. Godal, P. G. Smith, and F. Modabber. 1999. A randomised, double-blind, controlled trial of a killed L. major vaccine plus BCG against zoonotic cutaneous leishmaniasis in Iran. Vaccine 17:466.

33.Bahar, K., Y. Dowlati, B. Shidani, M. H. Alimohammadian, A. Khamesipour, S. Ehsasi, R. Hashemi-Fesharki, S. Ale-Agha, and F. Modabber. 1996. Comparative safety and immunogenicity trial of two killed *Leishmania major* vaccines with or without BCG in human volunteers. Clin. Dermatol. 14:489.

34.Connell, N. D., E. Medina-Acosta, W. R. McMaster, B. R. Bloom, and D. G. Russell. 1993. Effective immunization against cutaneous leishmaniasis with recombinant bacille Calmette-Guerin expressing the *Leishmania* surface proteinase gp63. Proc. Natl. Acad. Sci. U. S. A 90:11473.

35.Abdelhak, S., H. Louzir, J. Timm, L. Blel, Z. Benlasfar, M. Lagranderie, M. Gheorghiu, K. Dellagi, and B. Gicquel. 1995. Recombinant BCG expressing the *leishmania* surface antigen Gp63 induces protective immunity against *Leishmania major* infection in BALB/c mice. Microbiology 141:1585.

36.Streit, J. A., T. J. Recker, J. E. Donelson, and M. E. Wilson. 2000. BCG expressing LCR1 of *Leishmania* chagasi induces protective immunity in susceptible mice. Exp. Parasitol. 94:33.

37.McMahon-Pratt, D., D. Rodriguez, J. R. Rodriguez, Y. Zhang, K. Manson, C. Bergman, L. Rivas, J. F. Rodriguez, K. L. Lohman, and N. H. Ruddle. 1993. Recombinant vaccinia viruses expressing GP46/M-2 protect against *leishmania* infection. Infect. Immun. 61:3351.

38.Yang, D. M., N. Fairweather, L. L. Button, W. R. McMaster, L. P. Kahl, and F. Y. Liew. 1990. Oral Salmonella typhimurium (AroA-) vaccine expressing a major leishmanial surface protein (gp63) preferentially induces T helper 1 cells and protective immunity against leishmaniasis. J. Immunol. 145:2281.

39.Xu, D., S. J. McSorley, S. N. Chatfield, G. Dougan, and F. Y. Liew. 1995. Protection against *Leishmania major* infection in genetically susceptible BALB/c mice by gp63 delivered orally in attenuated Salmonella typhimurium (AroA- AroD-). Immunology 85:1.

40.McSorley, S. J., D. Xu, and F. Y. Liew. 1997. Vaccine efficacy of Salmonella strains expressing glycoprotein 63 with different promoters. Infect. Immun. 65:171.

41.Gonzalez, C. R., F. R. Noriega, S. Huerta, A. Santiago, M. Vega, J. Paniagua, V. Ortiz-Navarrete, A. Isibasi, and M. M. Levine. 1998. Immunogenicity of a Salmonella typhi CVD 908 candidate vaccine strain expressing the major surface protein gp63 of *Leishmania mexicana mexicana*. Vaccine 16:1043.

42.Afonso, L. C., T. M. Scharton, L. Q. Vieira, M. Wysocka, G. Trinchieri, and P. Scott. 1994. The adjuvant effect of interleukin-12 in a vaccine against *Leishmania major*. Science 263:235

43.Kenney, R. T., D. L. Sacks, J. P. Sypek, L. Vilela, A. A. Gam, and K. Evans-Davis. 1999. Protective immunity using recombinant human IL-12 and alum as adjuvants in a primate model of cutaneous leishmaniasis. J. Immunol. 163:4481.

44.Campos-Neto, A., R. Porrozzi, K. Greeson, R. N. Coler, J. R. Webb, Y. A. Seiky, S. G. Reed, and G. Grimaldi, Jr. 2001. Protection against cutaneous leishmaniasis induced by recombinant antigens in murine and nonhuman primate models of the human disease. Infect. Immun. 69:4103.

45.Donnelly, J. J., J. B. Ulmer, J. W. Shiver, and M. A. Liu. 1997. DNA vaccines. Annu. Rev. Immunol. 15:617.

46.Shedlock, D. J. and D. B. Weiner. 2000. DNA vaccination: antigen presentation and the induction of immunity. J. Leukoc. Biol. 68:793.

47.Oliveira, S. C., G. M. Rosinha, C. F. de Brito, C. T. Fonseca, R. R. Afonso, M. C. Costa, A. M. Goes, E. L. Rech, and V. Azevedo. 1999. Immunological properties of gene vaccines delivered by different routes. Braz. J. Med. Biol. Res. 32:207.

48.Feltquate, D. M., S. Heaney, R. G. Webster, and H. L. Robinson. 1997. Different T helper cell types and antibody isotypes generated by saline and gene gun DNA immunization. J. Immunol. 158:2278.

49.Raz, E., H. Tighe, Y. Sato, M. Corr, J. A. Dudler, M. Roman, S. L. Swain, H. L. Spiegelberg, and D. A. Carson. 1996. Preferential induction of a Th1 immune response and inhibition of specific IgE antibody formation by plasmid DNA immunization. Proc. Natl. Acad. Sci. U. S. A 93:5141.

50.Pertmer, T. M., T. R. Roberts, and J. R. Haynes. 1996. Influenza virus nucleoprotein-specific immunoglobulin G subclass and cytokine responses elicited by DNA vaccination are dependent on the route of vector DNA delivery. J. Virol. 70:6119.

51. Asakura, Y., L. J. Liu, N. Shono, J. Hinkula, A. Kjerrstrom, I. Aoki, K. Okuda, B. Wahren, and J. Fukushima. 2000. Th1-biased immune responses induced by DNA-based immunizations are mediated via action on professional antigen-presenting cells to up-regulate IL-12 production. Clin. Exp. Immunol. 119:130.

52. Kowalczyk, D. W. and H. C. Ertl. 1999. Immune responses to DNA vaccines. Cell Mol. Life Sci. 55:751.

53.Alarcon, J. B., G. W. Waine, and D. P. McManus. 1999. DNA vaccines: technology and application as anti-parasite and anti-microbial agents. Adv. Parasitol. 42:343.

54.Liu, M. A., T. M. Fu, J. J. Donnelly, M. J. Caulfield, and J. B. Ulmer. 1998. DNA vaccines. Mechanisms for generation of immune responses. Adv. Exp. Med. Biol. 452:187.

55.Wong, J. P., M. A. Zabielski, F. L. Schmaltz, G. G. Brownlee, L. A. Bussey, K. Marshall, T. Borralho, and L. P. Nagata. 2001. DNA vaccination against respiratory influenza virus infection. Vaccine 19:2461.

56.Liu, M. A., W. McClements, J. B. Ulmer, J. Shiver, and J. Donnelly. 1997. Immunization of non-human primates with DNA vaccines. Vaccine 15:909.

57.Lodmell, D. L. and L. C. Ewalt. 2000. Rabies vaccination: comparison of neutralizing antibody responses after priming and boosting with different combinations of DNA, inactivated virus, or recombinant vaccinia virus vaccines. Vaccine 18:2394.

58.Schultz, J., G. Dollenmaier, and K. Molling. 2000. Update on antiviral DNA vaccine research (1998-2000). Intervirology 43:197.

59.Kim, J. J. and D. B. Weiner. 1999. Development of multicomponent DNA vaccination strategies against HIV. Curr. Opin. Mol. Ther. 1:43.

60.Letvin, N. L. 1998. Progress in the development of an HIV-1 vaccine. Science 280:1875.

61.Coler, R. N., A. Campos-Neto, P. Ovendale, F. H. Day, S. P. Fling, L. Zhu, N. Serbina, J. L. Flynn, S. G. Reed, and M. R. Alderson. 2001. Vaccination with the T cell antigen Mtb 8.4 protects against challenge with Mycobacterium tuberculosis. J. Immunol. 166:6227.

62.Skeiky, Y. A., P. J. Ovendale, S. Jen, M. R. Alderson, D. C. Dillon, S. Smith, C. B. Wilson, I. M. Orme, S. G. Reed, and A. Campos-Neto. 2000. T cell expression cloning of a Mycobacterium tuberculosis gene encoding a protective antigen associated with the early control of infection. J. Immunol. 165:7140.

63.Lowrie, D. B., R. E. Tascon, V. L. Bonato, V. M. Lima, L. H. Faccioli, E. Stavropoulos, M. J. Colston, R. G. Hewinson, K. Moelling, and C. L. Silva. 1999. Therapy of tuberculosis in mice by DNA vaccination. Nature 400:269.

64.Lai, W. C., S. P. Pakes, K. Ren, Y. S. Lu, and M. Bennett. 1997. Therapeutic effect of DNA immunization of genetically susceptible mice infected with virulent Mycoplasma pulmonis. J. Immunol. 158:2513.

65.Wang, R., J. Epstein, F. M. Baraceros, E. J. Gorak, Y. Charoenvit, D. J. Carucci, R. C. Hedstrom, N. Rahardjo, T. Gay, P. Hobart, R. Stout, T. R. Jones, T. L. Richie, S. E. Parker, D. L. Doolan, J. Norman, and S. L. Hoffman. 2001. Induction of CD4+ T cell-dependent CD8+ type 1 responses in humans by a malaria DNA vaccine. Proc. Natl. Acad. Sci. U. S. A.

66.Daubersies, P., A. W. Thomas, P. Millet, K. Brahimi, J. A. Langermans, B. Ollomo, L. BenMohamed, B. Slierendregt, W. Eling, A. Van Belkum, G. Dubreuil, J. F. Meis, C. Guerin-Marchand, S. Cayphas, J. Cohen, H. Gras-Masse, and P. Druilhe. 2000. Protection against Plasmodium falciparum malaria in chimpanzees by immunization with the conserved pre-erythrocytic liver-stage antigen 3. Nat. Med. 6:1258.

67.Mendez, S., S. Gurunathan, S. Kamhawi, Y. Belkaid, M. A. Moga, Y. A. Skeiky, A. Campos-Neto, S. Reed, R. A. Seder, and D. Sacks. 2001. The potency and durability of DNA- and protein-based vaccines against Leishmania major evaluated using low-dose, intradermal challenge. J. Immunol. 166:5122.

68.Fragaki, K., I. Suffia, B. Ferrua, D. Rousseau, Y. Le Fichoux, and J. Kubar. 2001. Immunisation with DNA encoding Leishmania infantum protein papLe22 decreases the frequency of parasitemic episodes in infected hamsters. Vaccine 19:1701.

69.Melby, P. C., G. B. Ogden, H. A. Flores, W. Zhao, C. Geldmacher, N. M. Biediger, S. K. Ahuja, J. Uranga, and M. Melendez. 2000. Identification of vaccine candidates for experimental visceral leishmaniasis by immunization with sequential fractions of a cDNA expression library. Infect. Immun. 68:5595.

70.Handman, E., A. H. Noormohammadi, J. M. Curtis, T. Baldwin, and A. Sjolander. 2000. Therapy of murine cutaneous leishmaniasis by DNA vaccination. Vaccine 18:3011.

71.Piedrafita, D., D. Xu, D. Hunter, R. A. Harrison, and F. Y. Liew. 1999. Protective immune responses induced by vaccination with an expression genomic library of Leishmania major. J. Immunol. 163:1467.

72.Walker, P. S., T. Scharton-Kersten, E. D. Rowton, U. Hengge, A. Bouloc, M. C. Udey, and J. C. Vogel. 1998. Genetic immunization with glycoprotein 63 cDNA results in a helper T cell type 1 immune response and protection in a murine model of leishmaniasis. Hum. Gene Ther. 9:1899.

73.Gurunathan, S., D. L. Sacks, D. R. Brown, S. L. Reiner, H. Charest, N. Glaichenhaus, and R. A. Seder. 1997. Vaccination with DNA encoding the immunodominant LACK parasite antigen confers protective immunity to mice infected with *Leishmania major*. J. Exp. Med. 186:1137.

74.Sepulveda, P., M. Hontebeyrie, P. Liegeard, A. Mascilli, and K. A. Norris. 2000. DNA-Based immunization with Trypanosoma cruzi complement regulatory protein elicits complement lytic antibodies and confers protection against Trypanosoma cruzi infection. Infect. Immun. 68:4986.

75.Wizel, B., N. Garg, and R. L. Tarleton. 1998. Vaccination with trypomastigote surface antigen 1-encoding plasmid DNA confers protection against lethal Trypanosoma cruzi infection. Infect. Immun. 66:5073.

76.Costa, F., G. Franchin, V. L. Pereira-Chioccola, M. Ribeirao, S. Schenkman, and M. M. Rodrigues. 1998. Immunization with a plasmid DNA containing the gene of trans-sialidase reduces Trypanosoma cruzi infection in mice. Vaccine 16:768.

77.Gurunathan, S., C. Y. Wu, B. L. Freidag, and R. A. Seder. 2000. DNA vaccines: a key for inducing long-term cellular immunity. Curr. Opin. Immunol. 12:442.

78.Stobie, L., S. Gurunathan, C. Prussin, D. L. Sacks, N. Glaichenhaus, C. Y. Wu, and R. A. Seder. 2000. The role of antigen and IL-12 in sustaining Th1 memory cells in vivo: IL-12 is required to maintain memory/effector Th1 cells sufficient to mediate protection to an infectious parasite challenge. Proc. Natl. Acad. Sci. U. S. A 97:8427.

79.Barouch, D. H., S. Santra, J. E. Schmitz, M. J. Kuroda, T. M. Fu, W. Wagner, M. Bilska, A. Craiu, X. X. Zheng, G. R. Krivulka, K. Beaudry, M. A. Lifton, C. E. Nickerson, W. L. Trigona, K. Punt, D. C. Freed, L. Guan, S. Dubey, D. Casimiro, A. Simon, M. E. Davies, M. Chastain, T. B. Strom, R. S. Gelman, D. C. Montefiori, M. G. Lewis, E. A. Emini, J. W. Shiver, and N. L. Letvin. 2000. Control of viremia and prevention of clinical AIDS in rhesus monkeys by cytokine-augmented DNA vaccination. Science 290:486.

80.Barouch, D. H., A. Craiu, M. J. Kuroda, J. E. Schmitz, X. X. Zheng, S. Santra, J. D. Frost, G. R. Krivulka, M. A. Lifton, C. L. Crabbs, G. Heidecker, H. C. Perry, M. E. Davies, H. Xie, C. E. Nickerson, T. D. Steenbeke, C. I. Lord, D. C. Montefiori, T. B. Strom, J. W. Shiver, M. G. Lewis, and N. L. Letvin. 2000. Augmentation of immune responses to HIV-1 and simian immunodeficiency virus DNA vaccines by IL-2/Ig plasmid administration in rhesus monkeys. Proc. Natl. Acad. Sci. U. S. A 97:4192.

81.Murphy, K. M. 1998. T lymphocyte differentiation in the periphery. Curr. Opin. Immunol. 10:226.

82.Trinchieri, G. 1995. Interleukin-12: a proinflammatory cytokine with immunoregulatory functions that bridge innate resistance and antigen-specific adaptive immunity. Annu. Rev. Immunol. 13:251.

83.Asnagli, H. and K. M. Murphy. 2001. Stability and commitment in T helper cell development. Curr. Opin. Immunol. 13:242.

84.Constant, S. L. and K. Bottomly. 1997. Induction of Th1 and Th2 CD4+ T cell responses: the alternative approaches. Annu. Rev. Immunol. 15:297.

85. Mosmann, T. R. and S. Sad. 1996. The expanding universe of T-cell subsets: Th1, Th2 and more. Immunol. Today 17:138.

86. Biron, C. A. and Gazzinelli, R. T. 1995. Effects of IL-12 on immune responses to microbial infections: a key mediator in regulating disease outcomeCurr. Opin. Immunol. 7: 485-496.

87. Kim, J. J., Nottingham, L. K., Tsai, A., Lee, D. J., Maguire, H. C., Oh, J., Dentchev, T., Manson, K. H., Wyand, M. S., Agadjanyan, M. G. Ugen KE, Weiner DB. 1999, Antigen-specific humoral and cellular immune responses can be modulated in rhesus macaques through the use of IFN-gamma, IL-12, or IL-18 gene adjuvants. J. Med. Primatol. 28: 214 - 223.

88. Trinchieri, G. 1997 Cytokines acting on or secreted by macrophages during intracellular infection (IL-10, IL-12, IFN-gamma) Curr. Opin. Hematol. 4: 59-66.

89. Krieg, A. M., A. K. Yi, and G. Hartmann. 1999. Mechanisms and therapeutic applications of immune stimulatory cpG DNA. Pharmacol. Ther. 84:113.

90.Davis, H. L. 2000. Use of CpG DNA for enhancing specific immune responses. Curr. Top. Microbiol. Immunol. 247:171.

92.Krieg, A. M. 2000. Immune effects and mechanisms of action of CpG motifs. Vaccine 19:618.

93.McCluskie, M. J., R. D. Weeratna, P. J. Payette, and H. L. Davis. 2001. The use of CpG DNA as a mucosal vaccine adjuvant. Curr. Opin. Investig. Drugs 2:35.

94.Zimmermann, S., O. Egeter, S. Hausmann, G. B. Lipford, M. Rocken, H. Wagner, and K. Heeg. 1998. CpG oligodeoxynucleotides trigger protective and curative Th1 responses in lethal murine leishmaniasis. J. Immunol. 160:3627.

95.Stacey, K. J. and J. M. Blackwell. 1999. Immunostimulatory DNA as an adjuvant in vaccination against Leishmania major. Infect. Immun. 67:3719.

96.Walker, P. S., T. Scharton-Kersten, A. M. Krieg, L. Love-Homan, E. D. Rowton, M. C. Udey, and J. C. Vogel. 1999. Immunostimulatory oligodeoxynucleotides promote protective immunity and provide systemic therapy for leishmaniasis via IL-12- and IFN-gamma-dependent mechanisms. Proc. Natl. Acad. Sci. U. S. A 96:6970.

97.Iho, S., T. Yamamoto, T. Takahashi, and S. Yamamoto. 1999. Oligodeoxynucleotides containing palindrome sequences with internal 5'-CpG-3' act directly on human NK and activated T cells to induce IFN-gamma production in vitro. J. Immunol. 163:3642.

98.Deml, L., R. Schirmbeck, J. Reimann, H. Wolf, and R. Wagner. 1999. Immunostimulatory CpG motifs trigger a T helper-1 immune response to human immunodeficiency virus type-1 (HIV-1) gp 160 envelope proteins. Clin. Chem. Lab Med. 37:199.

99.Sato, Y., M. Roman, H. Tighe, D. Lee, M. Corr, M. D. Nguyen, G. J. Silverman, M. Lotz, D. A. Carson, and E. Raz. 1996. Immunostimulatory DNA sequences necessary for effective intradermal gene immunization. Science 273:352.

100.Vogel, F. R. and M. F. Powell. 1995. A compedium of vaccine adjuvants excipients. In Vaccine Design, The Subunit and Adjuvant Approach. Michael F.Powell and Mark J.Newman, eds. Plenum Press, New York and London, pp. 141-228.

101.Edelman, R. 1997. Adjuvants for the future. In New Genaration Vaccines. Myron M.Levine, Graeme C.Woodrow, James B.Kaper, and Gary S.Cobon, eds. Marcel Dekker, Inc., New York, Basel, Hong Kong, pp. 173-192.

102.Titus, R. G., F. J. Gueiros-Filho, L. A. de Freitas, and S. M. Beverley. 1995. Development of a safe live Leishmania vaccine line by gene replacement. Proc. Natl. Acad. Sci. U. S. A 92:10267.

103.Veras, P., C. Brodskyn, F. Balestieri, L. Freitas, A. Ramos, A. Queiroz, A. Barral, S. Beverley, and M. Barral-Netto. 1999. A dhfr-ts- Leishmania major knockout mutant cross-protects against Leishmania amazonensis. Mem. Inst. Oswaldo Cruz 94:491.

104.Belkaid, Y., S. Kamhawi, G. Modi, J. Valenzuela, N. Noben-Trauth, E. Rowton, J. Ribeiro, and D. L. Sacks. 1998. Development of a natural model of cutaneous leishmaniasis: powerful effects of vector saliva and saliva preexposure on the long-term outcome of Leishmania major infection in the mouse ear dermis. J. Exp. Med. 188:1941.

105.Kamhawi, S., Y. Belkaid, G. Modi, E. Rowton, and D. Sacks. 2000. Protection against cutaneous leishmaniasis resulting from bites of uninfected sand flies. Science 290:1351.

106.Sacks, D. L. 2001. Leishmania-sand fly interactions controlling species-specific vector competence. Cell Microbiol. 3:189.

107.Belkaid, Y., J. G. Valenzuela, S. Kamhawi, E. Rowton, D. L. Sacks, and J. M. Ribeiro. 2000. Delayed-type hypersensitivity to Phlebotomus papatasi sand fly bite: An adaptive response induced by the fly? Proc. Natl. Acad. Sci. U. S. A 97:6704.

108.Valenzuela, J. G., Y. Belkaid, M. K. Garfield, S. Mendez, S. Kamhawi, E. D. Rowton, D. L. Sacks, and J. M. Ribeiro. 2001. Toward a defined anti-*Leishmania* vaccine targeting vector antigens: characterization of a protective salivary protein. J. Exp. Med. 194:331.

109.Aebischer, T., M. Wolfram, S. I. Patzer, T. Ilg, M. Wiese, and P. Overath. 2000. Subunit vaccination of mice against new world cutaneous leishmaniasis: comparison of three proteins expressed in amastigotes and six adjuvants. Infect. Immun. 68:1328.

110.Webb, J. R., A. Campos-Neto, P. J. Ovendale, T. I. Martin, E. J. Stromberg, R. Badaro, and S. G. Reed. 1998. Human and murine immune responses to a novel *Leishmania major* recombinant protein encoded by members of a multicopy gene family. Infect. Immun. 66:3279.

111.Soong, L., S. M. Duboise, P. Kima, and D. McMahon-Pratt. 1995. *Leishmania pifanoi* amastigote antigens protect mice against cutaneous leishmaniasis. Infect. Immun. 63:3559.

112.Scott, P. 1991. IFN-gamma modulates the early development of Th1 and Th2 responses in a murine model of cutaneous leishmaniasis. J. Immunol. 147:3149.

113.Champsi, J. and D. McMahon-Pratt. 1988. Membrane glycoprotein M-2 protects against *Leishmania* amazonensis infection. Infect. Immun. 56:3272.

114. Mougneau, E., F. Altare, A. E. Wakil, S. Zheng, T. Coppola, Z. E. Wang, R. Waldmann, R. M. Locksley, and N. Glaichenhaus. 1995. Expression cloning of a protective *Leishmania* antigen. Science 268:563.

115.Campos-Neto, A., L. Soong, J. L. Cordova, D. Sant'Angelo, Y. A. Skeiky, N. H. Ruddle, S. G. Reed, C. Janeway, Jr., and D. McMahon-Pratt. 1995. Cloning and expression of a *Leishmania donovani* gene instructed by a peptide isolated from major histocompatibility complex class II molecules of infected macrophages. J. Exp. Med. 182:1423.

116.Julia, V., M. Rassoulzadegan, and N. Glaichenhaus. 1996. Resistance to *Leishmania major* induced by tolerance to a single antigen. Science 274:421.

117.Amaral, V., C. Pirmez, A. Goncalves, V. Ferreira, and G. Grimaldi, Jr. 2000. Cell populations in lesions of cutaneous leishmaniasis of *Leishmania* (L.) amazonensis. Mem. Inst. Oswaldo Cruz 95:209.

118.Amaral, V. F., V. A. Ransatto, F. Conceicao-Silva, E. Molinaro, V. Ferreira, S. G. Coutinho, D. McMahon-Pratt, and G. Grimaldi, Jr. 1996. *Leishmania* amazonensis: the Asian rhesus macaques (Macaca mulatta) as an experimental model for study of cutaneous leishmaniasis. Exp. Parasitol. 82:34.

119.Baba, T. W., V. Liska, R. Hofmann-Lehmann, J. Vlasak, W. Xu, S. Ayehunie, L. A. Cavacini, M. R. Posner, H. Katinger, G. Stiegler, B. J. Bernacky, T. A. Rizvi, R. Schmidt, L. R. Hill, M. E. Keeling, Y. Lu, J. E. Wright, T. C. Chou, and R. M. Ruprecht. 2000. Human neutralizing monoclonal antibodies of the IgG1 subtype protect against mucosal simian-human immunodeficiency virus infection. Nat. Med. 6:200.

120.Githure, J. I., G. D. Reid, A. A. Binhazim, C. O. Anjili, A. M. Shatry, and L. D. Hendricks. 1987. *Leishmania major*: the suitability of East African nonhuman primates as animal models for cutaneous leishmaniasis. Exp. Parasitol. 64:438.

121.Githure, J. I., A. M. Shatry, R. Tarara, J. D. Chulay, M. A. Suleman, C. N. Chunge, and J. G. Else. 1986. The suitability of East African primates as animal models of visceral leishmaniasis. Trans. R. Soc. Trop. Med. Hyg. 80:575.

122.Walsh, G. P., E. V. Tan, E. C. dela Cruz, R. M. Abalos, L. G. Villahermosa, L. J. Young, R. V. Cellona, J. B. Nazareno, and M. A. Horwitz. 1996. The Philippine cynomolgus monkey (Macaca fasicularis) provides a new nonhuman primate model of tuberculosis that resembles human disease. Nat. Med. 2:430.

123.Ulrich, J. T. and Myers K.R. 1995. Monophosphoryl lypid A as an adjuvant. In Vaccine Design: The Subunit Adjuvant Aprroach. Michael F.Powell and Mark J.Newman, eds. Plenum Press, New York, pp. 495-524.

INDEX

DATE DUE

JAN 0 8 2005